# MODAL HOMOTOPY TYPE THEORY

# Modal Homotopy Type Theory

*The Prospect of a New Logic for Philosophy*

DAVID CORFIELD

OXFORD
UNIVERSITY PRESS

# OXFORD
UNIVERSITY PRESS

Great Clarendon Street, Oxford, OX2 6DP,
United Kingdom

Oxford University Press is a department of the University of Oxford.
It furthers the University's objective of excellence in research, scholarship,
and education by publishing worldwide. Oxford is a registered trade mark of
Oxford University Press in the UK and in certain other countries

First Edition published in 2020

Impression: 4

Published in the United States of America by Oxford University Press
198 Madison Avenue, New York, NY 10016, United States of America

British Library Cataloguing in Publication Data
Data available

Library of Congress Control Number: 2019953288

ISBN 978–0–19–885340–4

DOI: 10.1093/oso/9780198853404.001.0001

Printed and bound by
CPI Group (UK) Ltd, Croydon, CR0 4YY

# PREFACE

*The old logic put thought in fetters, while the new logic gives it wings. It has, in my opinion, introduced the same kind of advance into philosophy as Galileo introduced into physics, making it possible at last to see what kinds of problems may be capable of solution, and what kinds are beyond human powers. And where a solution appears possible, the new logic provides a method which enables us to obtain results that do not merely embody personal idiosyncrasies, but must command the assent of all who are competent to form an opinion.*

(Bertrand Russell, *Logic As The Essence Of Philosophy*, 1914)

Bertrand Russell, a little over a century ago, promised a great future in philosophy for the new logic devised by Frege and Peano. The first-order predicate logic that emerged from their work, 'by an analysis of mathematics' (Russell 1914, p. 50), has certainly spread far and wide in the anglophone philosophical world. But now in the early years of the twenty-first century, a new 'new logic' has appeared, a new foundation for mathematics. It arose through a philosopher-logician, Per Martin-Löf, rethinking the composite nature of mathematical judgement and formulating his ideas in the shape of a theory of *types*. Currents from this type theory blended with others deriving from another radical reconceptualization of the foundations of mathematics, this time in the shape of *category theory*, with its emphasis on structure and transformation. The tight relationship between type theory and category had long been studied, not least in computer science departments, but the meshing of these currents was propelled to take on the form of our new logic by the insights of one of the finest mathematicians of the past few decades, Vladimir Voevodsky, who was looking to develop proof assistants capable of dealing with modern mathematics (Voevodsky 2014). This new logic allows advanced *homotopical* concepts to be directly constructed and manipulated; in particular, the mathematical concepts used in current physics. The name of this wonder-language is *homotopy type theory*. In this book I shall be arguing that *philosophy* should look to *homotopy type theory* and variants, in particular what is known as *modal* homotopy type theory, as its formal language of choice.

'Plain' homotopy type theory, familiarly known as 'HoTT', has very recently appeared (UFP 2014) as a contender to challenge set theory's long-standing claim to act as the primary foundational language for mathematics. Yet more recently, it has been proposed that to this new language there be added extra resources, in the form of what are called *modalities*, in order to express concepts such as continuity and smoothness. Taken together, I will refer to languages in this family as 'modal homotopy type theory', or 'modal HoTT' for short. Already a large body of results from quantum gauge field theory has been written up in modal HoTT (Schreiber 2013, 2014a, forthcoming). Even so, I do not intend this book primarily as a contribution to the philosophical study of mathematics, and certainly not of physics. Indeed, an exploration of what modal HoTT can bring to a philosophical treatment of physics will have to wait for another book. Regarding mathematics, since set theory is formulated as an axiomatic system built on top of a base of first-order logic, and since the major part of the debate surrounding axiom choice is only directly relevant to mathematics, then to treat set theory philosophically is inevitably to look to speak to the philosophy of mathematics. The component which is first-order logic may be separated out and its particular relevance for philosophy then studied. HoTT by contrast comes as a whole package, with its logic integrated. From the perspective of HoTT, to apply its 'logic' in philosophy is already to apply the whole calculus. Similarly, we cannot separate out naturally a modal logic from modal HoTT, so we will need to think through applications of the whole calculus to philosophy. If Russell could compare the introduction of his new logic into philosophy with that of mathematics into physics, now we can consider the prospect of the new 'new logic' being introduced simultaneously into all three disciplines.

In view of the prominence accorded to logic in current anglophone philosophy, anywhere this philosophy has seen fit to employ what it takes as its standard logical tools—in the philosophy of language, in metaphysics or wherever—there is room to explore whether modal HoTT fares better. Over the course of the book, I aim to convince the reader that it does. Many philosophers throughout the twentieth century expressed their serious misgivings about the use of formal tools in philosophy, illustrated frequently by apparent failings of first-order logic to capture the nuances of natural language. I hope that I can give sufficient reason here to cast doubt on their pessimistic conclusions.

My promotion of modal HoTT is clearly, then, no minor proposal. We normally train our students in philosophy in propositional and then first-order logic. Those wishing to study metaphysics will probably also learn some modal logic. If I am right, all of this should change. New lectures and textbooks will have to be written, and a large-scale retraining exercise begin. But not only are the tools of the trade to change, the whole disciplinary landscape of philosophy will have to alter too. We have here a language which at last serves cutting-edge mathematics well, and consequently so too the theoretical physics which relies upon it. At the same time, I maintain, it makes a much better fist of capturing the structure of natural language. A radical reconfiguring of the relations of these domains to that of metaphysics is therefore to be expected.

Chapters 2 to 4 will see us build up the component parts of modal HoTT. In Chapter 2 we motivate and deploy type theory, and in particular the dependent version. Chapter 3 then explains why people have looked to represent more subtle notions of identity. Here we see in particular how the 'Ho' in HoTT is derived. Chapter 4 then introduces modalities to the type theory. This is followed up by Chap. 5, which illustrates what a particular variant of modal HoTT—the *differential cohesive* variety—can bring to the philosophy of geometry.

I begin, however, in Chap. 1 with an introductory survey of the kinds of thinking that have motivated the developments we will be covering through this book. I do so via an account of how I arrived at such a radical point of view as to wish to replace philosophy's logic. There will be allusion made to constructions that will be dealt with in closer detail in later chapters, but I hope the broadbrush picture given will help to orient readers. On the other hand, it would be quite possible for the reader to set out from Chap. 2.

The authors of the HoTT book can warn us that, because of its youth,

> This book should be regarded as a 'snapshot' of just one portion of the field, taken at the time it was written, rather than a polished exposition of a completed edifice. (UFP 2014, p. 1)

All the more, then, should such a warning apply to a book such as this. Alongside some of my own writings (Corfield 2017a, 2017b) there has been some early philosophical exploration of HoTT (see, for example, Awodey 2014, Ladyman and Presnell 2015, 2016, 2017, Tsementzis 2017, Walsh 2017), but, as I write, nothing concerning modal HoTT beyond what is included in an article of mine (Corfield 2017c). Even after a book-length treatment, I will have only scratched the surface of the philosophical relevance of modal HoTT. This is not just because I am only beginning here the task of thinking through what modal HoTT might be able to do for philosophy. It is also that the heuristic power of the HoTT programme has plenty of steam left for it to move in unforeseen ways. In view of the rapidity of developments in the field, we see further useful variants of type theory in the pipeline. Ideas here include: a *cubical* version, bringing the prospect of better computational properties; a *directed* version, which should encode one-way irreversible processes; and a *linear* version, which should be well-suited for quantum physics and stable homotopy theory. There are also *two-level* versions, which provide the means to define certain infinite structures, and a notion of 'type 2-theory' to describe higher structures. As I write these words, modal dependent type theory itself is being reformulated as 'dependent type 2-theory'. But no matter. Even if in a few years' time the consensus has shifted on the best way to frame the variety of type theories, taking efforts now to learn about some currently standard modal HoTT constructions and their applications will not be time wasted, since some form of this logical calculus is here to stay.

It is important to note that naming conventions have not been definitively settled yet. Some people have looked to distinguish *Univalent Foundations* from HoTT, but for the duration of this book, I shall be working on the understanding that HoTT is a variety of intensional Martin-Löf dependent type theory with higher inductive types and satisfying the univalence axiom. I shall be referring to all of these ingredients in the course of the book as we need them.[1]

Portions of the contents of two chapters of this book have appeared elsewhere. Chapter 3 is a reworking of a rather clumsily named paper of mine, 'Expressing "The Structure Of" in Homotopy Type Theory' (Corfield 2017a). In the setting of a book I now have the space to include considerably more detail on important ingredients of HoTT. Similarly, Chap. 5 provides an opportunity to recast my article 'Reviving the Philosophy of Geometry' (Corfield 2017c) in the context of a more elaborate discussion of the current situation in geometry, and with the background prepared by the preceding chapters, I have the opportunity to enter into much more detail about the geometric modalities involved.

---

[1] This is in line with the 'HoTT Book' (UFP 2014), which provides the most detailed account of the system in a reasonably accessible way. See also Shulman (2017) for an excellent introduction.

The pleasure I have taken in my research path over many years has been greatly enhanced by interacting with some of the people responsible for the construction of modal HoTT. In particular, over the past ten years I have been a steering committee member of the wonderful nLab wiki and a participant in its discussion forum, the nForum. The generosity of the contributors there, including two of the pivotal figures in the development of modal HoTT, Mike Shulman and Urs Schreiber, never ceases to astound me. It is a genuine privilege to witness the foundations of mathematics and physics being developed before your own eyes. Many of the ideas contained in this book have come from them, or been developed in discussions with them. For much more exposition about many of the concepts and broader outlooks treated in this book, the nLab is an invaluable resource to go alongside more conventional publications, such as the HoTT book (UFP 2014). I will provide additional pointers to these and other resources in the Further Reading section towards the end of the book. The nLab was spun off from a blog which Urs, Mike, I and others run, the n-Category Café. Its founding mission when Urs and I set it up with John Baez was to explore what higher category theory might mean for mathematics, physics and philosophy. We always operated under the firmly held tenet that careful explanation could bring clarity even to the most advanced concepts. I have endeavoured to adhere to this ideal. New contributors/editors to the nLab are always welcome. Visit the nForum for advice about how to get started.

In addition to my nLab coworkers, I should like to thank various members of audiences which have heard me: in particular, Jon Williamson, Colin McLarty, Zhaohui Luo and my doctoral student, Gavin Thomson.

I dedicate this book with much love to my mother and to the memory of my father.

# CONTENTS

# 1

## A Path to a New Logic

### 1.1 First Encounters

It is striking, although I imagine not a terribly uncommon experience, when you come to realize through the course of a life of intellectual enquiry that you haven't moved so very far from where you first began. Nearly thirty years ago I came to philosophy from mathematics, full of zeal to discover what should be said about the richness of the latter's concepts. I could see that beautiful thematic ideas were to be found manifesting themselves across the historic, yet frequently breached, boundaries regimenting its branches, and I pored over the writings of the philosophical master of these *supra*-mathematical ideas, Albert Lautman. I was convinced that a formal language I had just begun to study, *category theory*, was much better suited than alternatives to speak to these ubiquitous concepts, and that Lautman could be seen as anticipating its invention.[1] This seemed to be the case even with mathematical logic, as Lambek and Scott (1986) had recently shown. *Adjoint functors*, those mainstays of category theory, appeared to be just as prevalent in logic as in the rest of mathematics, something William Lawvere had demonstrated a couple of decades earlier (Lawvere 1969). In fact, I agreed wholeheartedly with Lautman that logic should be seen as of a piece with mathematics, and thought to look to category theory's treatment of logic to explain this to the philosophical community I was hoping to join.

Consequently, my Masters thesis addressed the relationship between two styles of semantics for intuitionistic logic, one couched in the traditional philosophical language of judgement and warrant, the other in spatial terms. Intuitionism, in this sense, has a long history stretching back to the sometimes mystical writings of the Dutch mathematician, L E J Brouwer, in the first decades of the twentieth century. Supposedly against his wishes, a logic had been extracted from these thoughts by Andrey Kolmogorov and Arend Heyting. Later, we could read philosophers, such as Michael Dummett, who argued that intuitionistic logic makes good sense of the distinctions in meaning Brouwer had unearthed, which were typically collapsed in classical hands.

---

[1] See the French mathematician Jean Dieudonné's comments to this effect in the preface to Lautman (2006), and also Corfield (2010).

*Modal Homotopy Type Theory: The Prospect of a New Logic for Philosophy.* David Corfield, Oxford University Press (2020). © David Corfield. DOI: 10.1093/oso/9780198853404.001.0001

In the meantime, however, a second style of semantics for this logic had emerged mid century from the work of McKinsey and Tarski, in a form which resembled somewhat the use of Venn diagrams to represent sets and their intersections. To illustrate something of the difference between these two kinds of semantics, consider the connective *implies* in propositional logic. On the one hand, there is the *proof-theoretic* interpretation whereby a proof of $A \rightarrow B$ is a means to transform any warrant for the truth of $A$ into a warrant for the truth of $B$. On the other hand, propositions in the *topological* semantics are interpreted spatially as interiors of domains, technically as *open subsets* of a topological space, denoted $A \mapsto (A)$. Then $A \rightarrow B$ is interpreted as the largest open set whose intersection with $(A)$ is contained in $(B)$. Dummett (1977) tried to assure us that the proof-theoretic interpretation was the proper one, and that the topological one was something of a fluke. While most readers will surely agree that the proof-theoretic account presents a more readily graspable meaning, at the time it struck me as implausible that the discovery of topological models for intuitionistic logic wasn't pointing us in some important direction, an indication again that logic ought to be considered as subsumed within mathematics. The tie between intuitionism and topology goes right back to the days of Brouwer, he himself being one of the founding fathers of topology from 1909, just a year after his first published criticism of the classical logical principle of excluded middle. It seemed to me at the time, then, that category theory could provide a resolution to the opposition between these twin semantics with its invention through the 1960s of *toposes*, categories which combine the logical and the spatial, and that this needed philosophical attention.[2]

During research for this work, I remember reading Göran Sundholm's resolution of the puzzle of the 'Donkey sentence' (Sundholm 1986) using a constructive type theory, a form of intuitionistic language, which resonated with Dummett's ideas. The original sentence is:

- If a farmer owns a donkey, then he beats it,

but Sundholm's point can be treated in terms of the simpler sentence

- If John owns a donkey, then he beats it.[3]

The problem here is that we expect there to be a compositional account of the meaning of this sentence as given by the sentence structure. At first glance, it appears that there is an existential quantifier involved in the antecedent of a conditional proposition, as signalled by the indefinite article. However, a beginner's attempt to use one is ill-formed, the final $x$ being unbound:

- $\exists x (Donkey(x) \& Owns(John, x)) \rightarrow Beats(John, x)$.

---

[2] Very recently in Shulman (forthcoming) we see an answer provided by modal homotopy type theory, as I explain in Chap. 5.

[3] I have heard the complaint that these are not the kinds of sentences ever likely to be uttered. Of course, it's easy to find examples with just the same structure which are more plausibly spoken: 'If a customer has purchased a faulty radio from my store, then she will receive a full refund for it.'

On the other hand, if the scope of ∃ is extended to the end of the sentence by a shift of parenthesis, then the new sentence means that there is some entity for which, if it is the case that it is a donkey owned by John, it is beaten by him. But then this is made true by a sheep owned by John or a donkey owned by Jane, whether beaten by John or not. Clearly, this is not the intended meaning.

The alternative in standard first-order logic is to rephrase the sentence as something like:

- Any donkey that John owns is beaten by him,

and then to render it formally as

- $\forall x(Donkey(x) \& Owns(John, x) \rightarrow Beats(John, x))$.

But can it really be the case that we comprehend such sentences by first performing this kind of radical transformation? Compositional accounts surely have the advantage of plausibility as concerns language acquisition and comprehension.

Sundholm showed how we could have our cake and eat it, a compositional and faithful account of the sentence using the resources of dependent type theory. Propositions in type theory are *types* whose elements are warrants for their truth. The idea, then, is to construct a type such that an element plays the role of justifying the Donkey sentence. Such an element would be of the kind that, when presented with a donkey together with a warrant that John owns it, would deliver a proof that he beats it. The resources of Martin-Löf's type theory allow precisely this. Its dependent sum (or sometimes *pair*) construction, $\sum$, provides a way to construct *paired* types, used here to form the type *Donkey owned by John,* $\sum x : Donkey(Owns(John, x))$. An element of this type has two components: a donkey and a proof of ownership by John. We can *project* out from any such pair to its components, using $p$ for the first and $q$ for the second.

Then the dependent product construction, $\prod$, provides a way to construct *function* types. An element of this type will send an element of the input type to one in the output type. The complete type for the proposition is then

- $\prod z : (\sum x : Donkey(Owns(John, x)))Beats(John, p(z)))$.

An element of this type takes any donkey owned by John to a proof that John beats that donkey. The 'it' of the Donkey sentence is represented by $p(z)$, the donkey component of the pair coupling the animal to a proof of its ownership by John. The product ($\prod$) at the beginning also confirms the thought that there's a form of hypothetical occurring, a 'whenever', and the sum ($\sum$) is a form of existential quantification.

All those years ago, I was not duly impressed with this contribution to the philosophy of language. What I cared about then was the promotion of category theory. Sundholm was relying on a formalism known after its inventor as *Martin-Löf dependent type theory,* and this had a semantics in what were called *locally cartesian closed categories,* the toposes mentioned above being a special case of these. What I might then have pursued was the ever-tighter embrace into which computer scientists, constructive type theorists and category theorists were entering. They were converging under the guidance of a vision dubbed by the computer scientist, Robert Harper, *computational trinitarianism*. Tongue in cheek, he glosses this position as follows:

> The central dogma of computational trinitarianism holds that Logic, Languages, and Categories are but three manifestations of one divine notion of computation. There is no preferred route to enlightenment: each aspect provides insights that comprise the experience of computation in our lives. (Harper 2011)

The serious point is that these three corners allow a literal 'triangulation' of the value of a new development put forward in any one of them. Any proposed construction in one field had better make good sense in the other two.

Constructive logic            Programming languages


Category theory

Taking category theory as the representative of mathematics, he continues:

> Imagine a world in which logic, programming, and mathematics are unified, in which every proof corresponds to a program, every program to a mapping, every mapping to a proof! Imagine a world in which *the code is the math*, in which there is no separation between the reasoning and the execution, no difference between the language of mathematics and the language of computing. Trinitarianism is the *central organizing principle* of a theory of computation that integrates, unifies, and enriches the language of logic, programming, and mathematics. It provides a *framework for discovery*, as well as analysis, of computational phenomena. An innovation in one aspect must have implications for the other; a good idea is a good idea, in whatever form it may arise. If an idea does not make good sense logically, categorially, and typically ... then it cannot be a manifestation of the divine. (Harper 2011)

There is a crucial lesson here—one that philosophers in particular have not always been the first to grasp. It is remarkably easy to convince yourself that you are doing good work when devising a formal system to capture some topic or other. Proper triangulation, where achievable, is enormously reassuring, then. If several (at least partially) independent sources agree that some construction or other looks right according to its own view of the matter, then there is a much greater chance that it will stand the test of time. It was with this lesson in mind that Sundholm had explained:

> In this manner, then, the type-theoretic abstractions suffice to solve the problem of the pronominal back-reference in [the Donkey sentence]. It should be noted that there is nothing *ad hoc* about the treatment, since all the notions used have been introduced for mathematical reasons in complete independence of the problem posed by [the Donkey sentence]. (Sundholm 1986, p. 503)

Indeed, the dependent sum and product constructions used there had already suggested themselves to Per Martin-Löf. As we shall see later, these constructions are also extremely natural from a category-theoretic perspective, namely, as *adjoints*.

Now, each of the corners of Harper's 'Trinity' comes with a very intricate history. It is important, then, to consider overlaps and differences of emphasis between them to understand the degree of their independence. Over many years, Martin-Löf has given us evolving versions of his constructive type theory. He achieved this by reaching back to Kant, Brentano, Frege and Husserl, that is, to traditions which take great care to articulate the judgemental aspects of logic, and then also to the school of Intuitionism in mathematics, founded by Brouwer and Heyting. It is not uncommon to hear people describe this latter movement, involving the denial of excluded middle and double negation elimination, as a failed revolution (see, for example, van Fraassen 2002, p. 239n8) which had been conclusively defeated in the world of mathematics at least as far back as the 1920s. Notably, one of the acknowledged world leaders of mathematics could say in that decade:

> Taking the principle of excluded middle from the mathematician would be the same, say, as proscribing the telescope to the astronomer or to the boxer the use of his fists. To prohibit existence statements and the principle of excluded middle is tantamount to relinquishing the science of mathematics altogether. (Hilbert 1927, p. 476)

However, constructivism refused to go away, and indeed computer science has very much taken it to heart. Any history of this convergence would dwell on the contribution of Andrei Markov, son of the identically named mathematician of Markov chain fame, whose brand of constructivism rested on the computational principles of recursion theory.

We can easily see why there should be a connection by considering an informal proof of the simple result that in a room with some people inside it, there must be a youngest person (being perhaps joint youngest):

- Assume not, so that we have a nonempty room where for everyone present there is a younger person.
- Then (without loss of generality) select someone in the room.
- By assumption, that person has someone younger than themselves. Choose one such person. Then again, that person has someone younger than them, and so on.
- In this way, we can create a sequence of people in the room whose ages grow ever younger of any length we like. Since there are only finitely many people present, any sufficiently long such sequence must have at least one person appear more than once. But then, by the transitivity of the 'younger than' relation, that someone is younger than themselves. Contradiction.
- So it can't possibly be that no one is the youngest. Therefore, someone must be the youngest.

This is a valid classical proof, but one which is needlessly non-constructive. Compare this with a method which actually allows you to lay your hands on the youngest:

- Place the room occupants in any order.
- Take the first person and compare their age with the next person.
- If older, switch to the new person. Otherwise, continue with the original person. Proceed to the third person and do likewise.
- Repeat until you reach the end of the line.
- The resulting person is youngest.

Of course, there are other ways to perform this task. I could arrange for the occupants to meet in tennis-tournament fashion, the younger of any pairing proceeding to the next round. Taking ages to be measured in a number of years, each of these are algorithms which effectively find a lowest natural number in a set (or perhaps multiset) of natural numbers.

From the constructive perspective, the needlessly non-constructive proof is still a proof of something, it's just not a proof of 'for a nonempty, finite (multi)set of natural numbers, there is a smallest member'. Instead, it is a proof of absurdity on assuming that in such a (multi)set no element is smallest. 'Proof of absurdity of...' is the form that constructive negation takes. The negation of a proposition, $\neg P$, is defined as $P \to \perp$, a proof of which when presented with a proof of $P$ yields a proof of absurdity. To argue by *reductio* as here, we would need the rule that $\neg\neg P$ implies $P$. Famously, in constructive logic we do not have this. We will see through the book that the constructive element within homotopy type theory (HoTT) will require us to be much more sensitive than is commonly the case in our use of negation.

Here in this case of a finite collection, our proof is evidently needlessly non-constructive, but there are cases involving infinitely large domains where I can avoid a *reductio* argument by invoking constructively valid principles. For instance, a proof that all natural numbers may be expressed as products of prime numbers can deploy the so-called *principle of strong induction* rather than the typical classical strategy of showing the absurdity of assuming there to be a least natural number that is not so expressible. On the other hand, there are cases where no valid constructive principle can step into the breach, such as when looking to establish that any real number is either less than zero, equal to zero or more than zero.

Clarifying the constructive-computer science connection, the so-called *Curry–Howard correspondence* associates constructive proofs of propositions to computer programs carrying out corresponding tasks. A proof is like an algorithm. A proof of, say, *A implies B* corresponds to a program which transforms an input of type *A* to an output of type *B*, in line with the interpretation by intuitionistic logic.

So if constructivism in logic and programming theory is integrally related, what, then, of the third corner, category theory? Well, one of its originators, Saunders Mac Lane, while a student in Göttingen in 1933, had an important brush with the kind of philosophy that informed Martin-Löf, including phenomenology through Oskar Becker (see McLarty 2007). Becker was someone whose philosophy inspired him to work out a formalism for intuitionistic logic, so there may be direct philosophical connections worth exploring. On the other hand, it is hard to overestimate the influence of William Lawvere on the category-theoretic outlook. The title of one of his papers, 'Taking Categories Seriously' (Lawvere 1986), beautifully encapsulates his outlook. To think category theory could provide a foundational language for mathematics (and physics) and to push through with this goal was extremely courageous, and is very far from being sufficiently recognized, even today.

Philosophically, Lawvere's points of nineteenth-century reference differ from Martin-Löf's, to include Hegel, Grassmann and Cantor. Very important historical work needs to be done to think through intellectual connections between his sources. Even so, it is quite possible to indicate quickly something of the commonality of category theory with constructivism in terms of certain structural observations. Let's consider again double negation, in particular the contrast between its introduction and its elimination rules. Constructive logicians are happy with the inference from the truth of $P$ to the truth of $\neg\neg P$, but not the other way around, $\neg\neg P$ to $P$. Rewriting these rules in constructive guise points us to the essential difference: $P\ true \vdash (P \to \bot) \to \bot\ true$ is an example of a very common mathematical construction. When a function, $f : A \to B$, is applied to an element, $a : A$, it results in an element $f(a) : B$. Thus, dually, any $a : A$ provides a means to turn any such function, $f$, into an element of $B$. There is a natural *application* pairing, $app : A \times B^A \to B$, which may be transformed (*curried*, to use the jargon) to an associated mapping, $A \to B^{B^A}$. Here an element $a$ of $A$ is sent to the map 'evaluate at $a$'.

Considering this phenomenon in the framework of propositions, and choosing $B$ as absurdity, we arrive at double negation introduction. This tallies with how the constructive logician explains things: if I have a proof of $P$, then consider the situation where I also have a proof of $\neg P$, or, in other words, a way of converting a proof of $P$ into a proof of absurdity. In that case, I would have a proof of absurdity. So a proof of $P$ is a means to transform a proof of $\neg P$ to absurdity, or, in other words, a proof of $\neg\neg P$.

The questions for the category theorist to ask, then, are: In what kinds of setting are these constructions possible? Where do we have for any two objects an object of functions between them? Where do we even have the ability to form a pair of items of different kinds, here a function and an argument? Answering these kinds of questions has been at the heart of categorical logic, and Lawvere has forged the way in extracting such common, deep principles. Even if people have heard of the structural set theory due to him known as the Elementary Theory of the Category of Sets (ETCS), or of the attempt to take a category of categories as foundational or even of his *elementary toposes*, one should not ignore Lawvere's success more generally in extracting the common essence of apparently diverse settings. In his paper 'Metric Spaces, Generalized Logic and Closed Categories' (Lawvere 1973), he speaks of a 'generalized logic' which makes apparent the commonality between ordinary categories and a kind of metric space, a space equipped with a distance. Categories collect together objects of a certain kind and relate them by transformations of the appropriate kind. Then for any two objects, $X$, $Y$, in a category, there is a 'hom-set', $Hom(X, Y)$, that is, the set of arrows or morphisms representing appropriate transformations between the objects. One of the fundamental rules of a category is the capacity to compose such arrows in the situation where the tail of one matches the head of another. So we require a composition map: $Hom(A, B) \times Hom(B, C) \to Hom(A, C)$. Now, Lawvere noted that this strongly resembles the triangle inequality, $d(x, y) + d(y, z) \geq d(x, z)$, which holds in metric spaces – it is always at least as far to go two sides around a triangle as along the third edge. He then showed that metric spaces can be seen as a variant of the former, where instead of the relationship between two objects being captured in a *set* of arrows, now the relationship between two points of space is determined by a *distance*, an element in the extended non-negative real numbers. So categories other than the category of sets may play the role of providing values for $Hom(A, B)$, and to do so they have to be equipped with additional structure. In the case here, a category which is enriched in the *monoidal* partially ordered set $([0, \infty], \geq)$ is a kind of metric space.

General enriched category theory allows a range of such monoidal categories to provide values for Hom objects, including the basic choice of the category of truth values. This is a category with only two objects, $\top$ and $\bot$. As required by a category, both objects have their identity arrow, representing the implications from *True* to *True* and *False* to *False*, and there is one further arrow from $\bot$ to $\top$, corresponding to the inference *False* to *True*. Evidently, we will not have an arrow in the other direction. In addition, we need a monoidal structure, here simply acting to multiply truth values by the ordinary propositional connective of conjunction. Now a category enriched in truth values amounts to a partially ordered set, a common mathematical structure.[4]

Returning to our search for an environment in which the pairing and unpairing of objects and the formation of function spaces are represented, category theory has determined the necessary features as that of being *cartesian* and being *closed*, respectively. These are very frequently encountered properties of categories. For a category to be cartesian it needs to possess a (finite) product structure, so in particular for any pair of objects, $A$ and $B$, a product, which is an object denoted $A \times B$, with projections to $A$ and $B$, such that any object with a pair of maps to $A$ and $B$ factors uniquely through the product. Being closed means that for any pair of objects, $B$ and $C$, there is an object $C^B$, such that $Hom(A \times B, C) \cong Hom(A, C^B)$. Combining these two features gives us the concept of a cartesian closed category. These are just the ingredients for the kind of setting where $A \mapsto B^{B^A}$ becomes a natural map.[5]

So now combining the idea of enrichment by truth values with the structure provided by being cartesian closed, we arrive at a category that behaves like a collection of propositions where arrows correspond to entailment. Consider a category of propositions with an arrow from $P$ to $Q$ whenever $Q$ follows from $P$. Then the product of $P$ and $Q$ is the conjunction $P\&Q$, and exponential objects are of the form $P \to Q$, hypothetical propositions. The cartesian closedness property of the category here amounts to $P \vdash Q \to R$ if and only if $P\&Q \vdash R$. Then, since we have the *modus ponens* entailment $P\&(P \to Q) \vdash Q$, so it follows that we have an entailment $P \vdash (P \to Q) \to Q$. In particular, when $Q$ is $\bot$, that is, falsity, then there is a natural map from $P$ to $\neg\neg P$.

To produce a map in the opposite direction, $\neg\neg P$ to $P$, as classical logic requires, will force us to add a much more specific structure to our categorical setting. There is a broad literature here which works with a cluster of relevant concepts involving *involution*, dagger categories, duals in vector spaces and Chu spaces, right up to star-autonomous categories, structures which have been made popular by categorical approaches to quantum mechanics (Abramsky and Coecke 2008). For our purposes, note that instead of the generic $Q$ that we

---

[4] Strictly speaking, one might say a *preorder*, where it is possible for $x \leq y \leq x$ without $x$ and $y$ being identical.
[5] We will revisit these structures in greater detail in Chap. 2.

find in the entailment $P \vdash (P \to Q) \to Q$, to provide an entailment in the other direction will involve a specific choice of object, known as a dualizing object (Corfield 2017d).

What I have sketched above is just the very beginning of a highly developed relationship between category theory and logic, both proof-theoretic and model-theoretic aspects, which has fed into theoretical computer science in a vast number of ways. We have a range of conditions which categorical settings may possess in accordance with different logical principles. A forensic scrutiny of the different pieces of a logical framework allows for subtle variants, including the typed lambda-calculus corresponding to cartesian closed categories, and versions of linear logic corresponding to varieties of monoidal category. You want conjunction, and there must be products. Weakening and deletion, and the category should be cartesian. A deduction theorem, and it should be cartesian closed. Meanwhile, reasoning in a higher-order constructive logic is possible in toposes. Finally, the subject of this book, HoTT, has been constructed with a view to providing the logical calculus for an $(\infty, 1)$-topos, a kind of higher category that has been devised by mathematicians in recent years. As a computationally salient language, proof assistants based on HoTT are flourishing. A new chapter for the *trinitarian* thesis is being written. I should now explain the need for higher categories.

## 1.2 Next Steps

Let me resume my own path which led me next to *groupoids* and then *higher-dimensional algebra*, vital ingredients for the HoTT outlook. As we shall see in Chap. 3, groupoids mark the third stage in a hierarchy which sets out with *propositions* and then *sets*. Perhaps the simplest way to think of them is as a common generalization of *equivalence relations* and of *groups*. If an equivalence relation divides a collection of entities into subcollections of mutually equivalent entities, a group can be thought of as collecting the ways in which a single entity is self-identical. So, we may divide, say, a population of people into equivalence classes of those individuals who have the same age. Between any two people the relation 'having the same age' is a yes–no matter. In particular, any person has the same age as herself. On the other hand, to think of groups in a similar light, consider a single object which is multiply related to itself. For instance, take a symmetric butterfly which looks identical to its mirror image, or take the twenty-four rotations of a cube which leave it invariant. The collections of such symmetries form groups.

In diagrammatic form, an equivalence relation corresponds to a set of elements arranged in separate clusters, where each pair in the same cluster is joined by a linking edge, including a reflexive loop at each point. In our same-aged people, the lines mark the equality of age. The corresponding diagram for a group of symmetries is a single point with a collection of loops corresponding to the group elements or symmetries. For the cube, then, we have a single point with twenty-four loops (along with a way to compose the corresponding transformations). There were several properly mathematical motivations for doing so, but even without these one might think to combine the concept of an equivalence relation with that of a group. The resulting concept is that of a groupoid. With this concept in hand, we have a way of representing a collection of entities, any pair of which may be inequivalent or equivalent in possibly multiple ways. For what seemed a very natural generalization of groups, the adoption of groupoids met with some considerable resistance in the mathematical community, something noted by prominent mathematicians such as Alain Connes. Consequently, groupoids made for an interesting case study, exploring the value judgements of mathematicians as to when a piece of concept stretching is essential, useful or useless, which I detailed in Corfield 2003, Chap. 9.

Researching for the paper that formed the basis for that chapter in the late 1990s, I came across the online writings of John Baez, one of the earliest mathematics expositors to use the internet in his column, 'This Week's Finds'. Interspersed throughout his essays could be found an account of an emerging way to deepen category theory. Now not only could we represent processes transforming some object to another, but also processes between processes, and so on. Later, in 2006, I teamed up with Baez and a third member, Urs Schreiber, to form a weblog, *The n-Category Café*, to discuss the relevance of higher categories for philosophy, mathematics and physics. One of the early explorations I prompted was to 'categorify' Euclidean geometry. Baez had written extensively on the advantage of working with the extra information present in categories that is lost when counting isomorphic entities as the same (see, for example, Baez and Dolan 2001). A simple tale of the rise of the natural numbers considers them to have been extracted from the identification of equinumerous finite sets. We say that the set of natural numbers is the *decategorification* of the category of finite sets, its equivalence classes generated by the relation of being isomorphic. Useful work could then be done to try to reverse this process by *categorifying* pieces of mathematics, that is, finding structures which when decategorified result in familiar pieces of mathematics. Decategorification is an algorithmic process, whereas the opposite direction may have none, one or several solutions.

My question, then, was this: If we are to see the arithmetic of the natural numbers as the shadow of operations on finite sets, why not see ordinary geometry as a shadow of some higher geometry? Just as Felix Klein had unified a range of different geometries with his *Erlangen Program* of the 1870s by taking a geometry to be the study of invariants of spaces under transformation by certain continuous groups, so we now went looking for *2-groups* to play this role, devices which could capture not only symmetries, but symmetries between symmetries. These 2-groups are examples of 2-groupoids, the next step up the HoTT hierarchy which comes after groupoids.[6]

For a snapshot of Baez's interests at that time, there is his wide-ranging article 'Physics, Topology, Logic and Computation: A Rosetta Stone', a piece he wrote with his student, Mike Stay (Baez and Stay 2010). The title alludes, of course, to the engraved stone which, bearing as it does a common text in three languages—hieroglyphics, demotic Egyptian and ancient Greek—allowed Champollion to decode the former in the early years of the nineteenth century. The Rosetta metaphor had already been used in mathematics by the French mathematician, André Weil, in a letter to his sister, the philosopher Simone Weil, describing the core ideas of his research and written while in prison in 1940 for desertion

---

[6] Without reaching fulfilment under my initial prompting, the idea has flourished in the form of higher Cartan geometry, as used recently in physics to approach M-theory, the theory proposed as unifying the five varieties of string theory. See Huerta, Sati, and Schreiber (2018, p. 5).

from the French army. Weil explained how he needed to translate between three languages (Riemannian surfaces, algebraic number fields and finite function fields) with the prospect of being able to translate a proof of one dialect's version of the Riemann hypothesis into a proof of the standard Riemann hypothesis.[7] This goal still hasn't been achieved many years later, although the ties between these dialects have been strengthened, in particular by the Langlands Program.

Baez and Stay are appealing, then, to perhaps a looser version of Harper's triangulation idea, but here with physics included, and topology appearing as a specific branch of mathematics:

> there is an extensive network of interlocking analogies between physics, topology, logic and computer science. They suggest that research in the area of common overlap is actually trying to build a new science: *a general science of systems and processes*. (Baez and Stay 2010, p. 97)

These analogies are summarized by Baez and Stay in the following table:

**Table 1.1** The Rosetta Stone.

| Category Theory | Physics | Topology | Logic | Computation |
| --- | --- | --- | --- | --- |
| object | system | manifold | proposition | data type |
| morphism | process | cobordism | proof | program |

Again, we find the Curry–Howard correspondence appearing, relating proofs to programs, and category theory tying these together in trinitarian fashion; however, now comparison is extended to include physical processes, mediating between initial and final states, and smooth oriented mathematical spaces, mediating between two parts of their boundary. In the case of the latter, natural extensions take in more complicated spaces of higher-dimensional manifolds with corners. This points to the need for higher categories:

> The Rosetta Stone we are describing concerns only the $n = 1$ column of the Periodic Table. So, it is probably just a fragment of a larger, still buried $n$-categorical Rosetta Stone. (p. 120)

In the case of physics, these higher dimensions had already started to be uncovered. Indeed, at *The n-Category Café*, Schreiber's interests combined with Baez's in the quest for a higher gauge theory (Baez and Schreiber 2005), with Urs's sights set firmly on string theory. If a particle traces out a one-dimensional path with the strength of the gauge field determining the force to which it is subjected, then a string sweeping out a two-dimensional surface should respond to something more intricate. So-called *second quantization* of string theory suggests yet higher-dimensional entities, termed $p$-branes, requiring higher Lie theory. Working out what constitutes gauge equivalence for these higher gauge fields requires

[7] See Chap. 4 of Corfield (2003).

the resources of higher category theory, since not only could fields be gauge equivalent, but there could be gauge of gauge equivalence, and so on. What we have here is a form of higher category theory, but one treating entities in which arrows of different levels have inverses up to equivalence at the next level. Here was an early appearance of our higher-groupoids, running right up the hierarchy to ∞-groupoids, for which there is no top level of arrow. Just as sets and sets with structures gather together to form categories, ∞-groupoids, possibly equipped with extra structure, gather together to form $(\infty, 1)$-categories. The category of sets itself forms a topos, whereas ∞-groupoids are gathered in an $(\infty, 1)$-category in the form of an $(\infty, 1)$-topos.[8]

At around the same time, the mathematicians André Joyal and Jacob Lurie showed that one could continue to use the constructions of category theory with very little modification in the setting of $(\infty, 1)$-categories, where instead of hom-*sets* between objects we have hom-*spaces*, or rather hom-*homotopy types*. Just as toposes are a kind of category with very pleasant features, in that, acting in ways similar to the category of sets, they support a rich logical structure, including the cartesian closedness property of the previous section, so there are some very nice $(\infty, 1)$-categories known as $(\infty, 1)$-toposes. Lurie has a book of over 1,000 pages, *Higher topos theory* (Lurie 2009a); this lays out the theory of these entities, which are being used in today's cutting-edge algebraic geometry.[9] Shortly after this, *homotopy type theory*, a descendant of one variety of Martin-Löf type theory, was proposed to perform the role of 'internal language' for $(\infty, 1)$-toposes.[10] An internal language for a kind of category is such that any valid reasoning carried out in that language may be interpreted in any category of that kind. This phenomenon has been exploited for many decades: such as when a constructive proof that a real matrix can be diagonalized is interpreted in the topos of sheaves over a space X. The interpretation then establishes that any symmetric matrix of real functions on X may be diagonalized (Scedrov 1986). This technique allows reasoning in a simple system to translate into a myriad of results when interpreted within categories supporting that reasoning. We shall revisit this theme again, especially in §6 of Chap. 5.

Now any thought that a category-theoretic foundation is some mere variant of a set-theoretic one becomes yet harder to maintain. Colin McLarty in his 'Uses and Abuses of the History of Topos Theory' (1990) argues convincingly that category theory never sought merely to explore variants of some category of sets, but rather was driven by a range of needs emerging in mainstream mathematics, especially algebraic topology and algebraic geometry. It is true that many commonly encountered categories are categories of structured sets and structure-preserving maps, such as groups and homomorphisms or topological spaces and continuous maps, but certainly by no means are they all of this form. For instance, in the category of sets and relations, an arrow is a relation, and this generalizes to a wide range of contexts in which *spans* or *correspondences* are gathered. These categories admit a duality, that is, any relation between A and B corresponds to one between

---

[8] Sometimes called a *higher* topos or an ∞-topos, although the latter term might be used for a different concept.

[9] We will meet briefly with some of Lurie's ideas in Chap. 5.

[10] Just as with ordinary toposes, there is a distinction between *Grothendieck* toposes and *elementary* toposes, the latter being a more general concept, so there is a similar distinction in the $(\infty, 1)$-versions. Suggestions for what ought to be the *elementary* form of these have been made, but the matter has not been fully resolved as yet.

$B$ and $A$, a feature not found in an ordinary set-function setting. Similarly, a cobordism category has arrows which are $n$-dimensional manifolds of some kind sitting between two parts of their $(n-1)$-dimensional boundaries, and any such manifold can be viewed in the opposite direction. But the further needs of current mathematics have given rise to a demand to understand how spaces relate to one another through further continuous spaces of mappings, and hence the drive to formulate $(\infty, 1)$-toposes with morphisms forming something more intricate than sets.

Over the past few years, variants of homotopy type theory have been proposed (cohesive, smooth, linear, directed) which aim to capture the mathematics needed to express further concepts of modern algebra and geometry, including that used in quantum gauge field theory and string theory. Where so much of contemporary mathematics, in particular that used in physics, is so very far from admitting a natural, purely set-theoretic reconstruction that is faithful to its ideas, now we have languages available which do not merely allow a formal rendering of its reasoning, but also provide assistance in concept formation right up at the front line of research. So we have a new 'logic', along with variants deriving from the heart of current mathematics and its application in fundamental physics, an earlier form of which, as we saw in the previous section, had achieved some notable success by Sundholm in compositionally representing natural language. As we shall see, this latter work marked the start of research which continues to this day, including the impressively thoroughgoing *Type-theoretical Grammar* by Aarne Ranta (1994). It would seem timely, then, for philosophical investigation to make sense of this logic, homotopy type theory, to see whether we might be able to forge advances in philosophy of language and metaphysics, such as were envisaged by the founding fathers of analytic philosophy. What objections could be raised to such a project?

## 1.3 Encounters with Ordinary Language Philosophy

To the extent that I have gained a reputation in philosophy, it has been for being anti-foundational or anti-formalist. A reader of Mancosu's editorial introduction to *The Philosophy of Mathematical Practice* (2008) or the review of my book, *Towards a Philosophy of Mathematics* (2003), by Paseau (2005) might easily have suspected that of all the people to be proposing an analytic philosophy Mark II, their number would not include myself. Well, I do confess to smiling when reading Gian-Carlo Rota, a professor of mathematics and philosophy at MIT, in his essay 'The Pernicious Influence of Mathematics upon Philosophy', likening analytic philosophers using formalisms to people trying to pay for their groceries with Monopoly money (Rota 1991, p. 169). And I remember reading Wittgenstein's warning with sympathy:

> 'Mathematical logic' has completely distorted the thinking of mathematicians and philosophers by declaring a superficial interpretation of the forms of our everyday language to be an analysis of the structures of facts. (Wittgenstein 1956, p. 156)

But if I appeared to Mancosu or Paseau to doubt the relevance of foundational formalisms for the philosophy of mathematics, and beyond, then this was a false impression. I was only ever properly opposed to what I took to be *inadequate* formalisms, although I seem to have more demanding criteria for what constitutes adequacy here. Many philosophers of mathematics rest content with a set theory which allows the translation *in principle* of any mathematical reasoning. However, it has struck me from my earliest encounters with category theory that, by comparison, set theory can only comprehend the palest shadow of the richness of mathematics, especially its geometric aspects. Criticism directed towards *some* formal methods does not equate to a condemnation of all ways. Why else would I have promoted higher category theory in the final chapter of my book?

As for uses of formalism more broadly in philosophy, arriving at Kent in 2007, I was exposed to a number of Rylean and Wittgensteinian thinkers keen to display the great elasticity of meaning in the ordinary use of language, and reluctant to acknowledge that a formal language was up to the task of making sense of our language with all of its subtle context dependence. From this point of view, even in the unlikely event that modal HoTT proved a good unifying language for mathematics, physics and computer science, there its range would likely end. Starting from the warning in the *Tractatus*:

> 4.002 Everyday language is a part of the human organism and is no less complicated than it. It is not humanly possible to gather immediately from it what the logic of language is,

and following Wittgenstein's disenchantment with logic, we find the ordinary language philosopher emphasizing 'motley' (*buntes Gemisch*) over regular order, so as to warn against the dangerous seductiveness of formalism. I particularly recall Peter Hacker, a visiting Professor at the time, asking an audience, 'What has the predicate calculus ever done for philosophy?'. Ordinary first-order logic does indeed seem inadequate for many philosophical purposes, but with our thesis that we should look to a richer formalism, modal HoTT, we may be able to put up stiffer resistance to the doubters.

For example, in my time I have heard the Wittgensteinian point out the foolishness of the philosopher lured by his or her logic into treating the 'is' of definition and the 'is' of description in the same way. The idea here is that there is a clear difference between *This is Prussian blue*, when I'm trying to teach you what this tone is by pointing to a sample, and *This is Prussian blue*, when I'm trying to inform you of the colour of some object you can't quite see. But why should there not be a way to distinguish these two kinds of *is*? Indeed, there is. At the heart of Martin-Löf type theory, there is a distinction between two equalities: *Judgemental* or *Definitional* on the one hand, and *Propositional* on the other. They rely on different syntax and different rules of use.

## *Definitional Equality*

This may be used in mathematics as a form of shorthand, where I might wish, say, to use a letter, $f$, to denote some specified function. In the case of colours here, I might use this form of equality to stipulate what a certain colour term means. Working with the type of colours, I will have established identity criteria for the various shades of colour. Perhaps a witness to the identity of two shades comes from the act of viewing instances together in good lighting conditions. So I can compare the colour of *this* with the colour of *that*, but, as any paint company will tell you, it is of course useful to provide standards. Their sample sheet is announcing definitional judgements of the form:

- The colour of this sample is Prussian blue.

Now I can apply the same constructions to these definitional equivalents, so that, for example, 'I like Prussian blue' has an equivalent meaning to 'I like the colour of this sample'.

## Propositional Equality

On the other hand, I may say to you 'This shirt is Prussian blue' in response to a question you put to me, or to contradict your own claim that 'This shirt is purple' made perhaps in a dim light, or with the garment in an opaque bag. Here I am saying that two terms denoting colours have identical reference:

- The colour of this shirt = the colour of the Prussian blue sample.

Were we to bring this shirt into proximity with the defining sample in good light, we could witness that they are the same colour. We say of these two terms that they are **propositionally equal**.

Naturally, these two forms of equality are distinguished within the formalism:

- *Prussian blue* $:\equiv$ *the colour of this sample* $:$ *Colour*.
- $p : Id_{Colour}(the\ colour\ of\ this\ shirt, Prussian\ blue)$.

The first of these is a *definitional* judgement. It is not subject to being proved, but is rather to be stipulated. The second is a *propositional* judgement that an identity type is inhabited. If the identity type *is* inhabited, in other words if we have established that there is an element of the type, then the corresponding proposition is true. We may use each of these judgements to derive further claims. In the case of the propositional judgement, assuming that 'to like' is behaving as one would hope on colours, we will be able to use $p$ to transform a warrant for the former proposition to one for the latter:

- $a : I\ like\ the\ colour\ of\ this\ shirt$.
- $p^*(a) : I\ like\ Prussian\ blue$.

On the other hand, definitional equality will allow for immediate substitutions, such as

- *I like Prussian blue* $\equiv$ *I like the colour of this sample* $:$ *Proposition*.

Syntactically, the behaviour is very different. The first equivalence is resulting from an element establishing identity; the second is arising from a stipulation. Of course, the status of something acting as a standard may change. An official colour chart that stipulates colours such as Prussian blue by means of a coloured square may be superseded. Then the shade of that obsolete coloured square becomes an empirical question. We see this phenomenon more clearly with shifts in the practice of metrology. The intricate history of modifications to the official standard for the metre length includes the decommissioning of a particular rod defining what it is to be a metre. Imagining it languishing in some warehouse today, we might wonder what its length might be as measured by current standards. This was not a question that could sensibly be posed during its heyday as the standard. There have, of course, been a great number of scientific factors driving these changes through to the current definition which is in terms of the distance travelled by light in a given length of time under certain conditions.

Returning to the verb 'to be', it is of course hugely overused in English. Indeed, yet a third 'is' should be distinguished. *Are there any colours? Yes, Prussian blue is a colour.* Again this is given a different treatment in the type theory

- *Prussian blue* : *Colour.*

Staying with colours, in the late 1920s, after writing the *Tractatus*, Wittgenstein was disturbed by the case of two propositions, 'This is green all over' and 'This is red all over', not being independent atoms, and yet not being identical or each other's negations. Where his thesis relied on descriptions of the world in terms of a collection of atomic propositions, composed of propositions whose truth values were mutually independent, it was not clear how to consider the two colour propositions above. Obviously, the truth of one excludes the truth of the other, and yet one is not the negation of the other. But then, this is so for any pair of different colours, and also for different sets of colours.

In view of this phenomenon and with no apparent prospect of resolving it within his system, for example, by further decomposition into atoms concerning light wavelength intervals, he famously turned away from his logical account and ascribed the incompatibility of colour claims to the grammatical rules of colour language. But, rather than aim one's fire at the poverty of a logic of atomic propositions in particular, why conclude so much the worse for logic in general? Why not move to a logic which allows one to represent a function, *'colours of'*, mapping from a type of object to something set-like, perhaps *Power set*(*Colour*), of discriminable tones? *Red* and *green* are both colours, and they're not the same. Distinctions within the target range of colours, or collections of colours, will be enough to force propositions concerning an entity's colour to behave in such a way as to contradict each other, just as statements reporting two different numbers resulting from counting the same collection contradict each other. For any particular concrete object sent to a singleton colour, for example {*red*}, we'd know it couldn't be sent to any different colour, for example {*green*}, {*blue*}, or even any set of colours, for example {*blue, green*}, {*red, green*}, the empty set, and so on.

The nature of *colours of* as a function to a subset of *Colour*, and so a set, integrates all the exclusion clauses for free. That such discrimination is made via the structure of the target of a map, the type of some chosen quantities, accounts for the lengthy steps science takes to make more precise colour ascriptions by appealing to the wavelength of light or, even more subtlely, a spectrogram. More precise discriminations can be made by the finer structure of the *codomain*, as category theorists term the target, so that, for instance, 'Wavelength of light$(x) = 650$ *nm*' and 'Wavelength of light$(x) = 450$ *nm*' exclude one another. The structure of the type of real numbers here entails that these predicates are incompatible. But rather than look to the structure of components of these sentences, Wittgenstein appealed to the grammar of our ordinary use of colour terms to explain these mutual incompatibilities.

Other philosophers have followed Wittgenstein in believing that logic cannot make sense of all of our inferences. Along with all those colour inferences—'X is red all over' implies 'X is not green all over'; 'X is red all over' implies 'X is not blue all over'—we have such incompatibilities as 'X is red all over' implies 'X is not blue and green'; 'X is red all over' implies 'X is not red and green'. Then, changing topic, we'd need 'if today is Wednesday, tomorrow is Thursday' for every day, 'Y is Austrian, so Y is European' for combinations of countries and continents, and so on over countless topics. For Wilfrid Sellars, reasoning of this kind is valid but in a *material* sense, according to a 'material rule of inference' (Sellars 1953, p. 313). Robert Brandom joins Sellars in his scepticism towards *enthymematic* reasoning being at play, with silently operating hypothetical statements. And indeed, if ordinary predicate logic is to represent the extent of our reasoning powers, this conclusion seems inevitable. There would have to be a tremendous number of implicit premises operating, the vast majority never considered before. My suggestion, however, is that a type theory such as HoTT can look to meet this challenge by showing in each case that the inference structure is supported by the rules of the type theory.

Let us, then, try another case. Remaining with colours, Sellars denies of the inference from

- The ball is red

to

- The ball is coloured

that we should consider there to be operating the implicit premise that

- Any red thing is coloured.

Instead, we assent to the correctness of the material inference due to our grasp of the concepts involved. It is only when such an inference comes in for scrutiny, perhaps through disagreement between reasoners, that the resources of our language allow us to consider its validity by *making explicit* the reasoning in a formal mode. The formal scheme of a piece of reasoning is located against the backdrop of communal practices of material inference.

On the other hand, the type theorist considering this inference will be led to consider how the propositions themselves are formed. Judgements as to their being propositions, and so types, must arise from pre-existing formation rules. Perhaps then, with these formation rules, along with further rules for the elements of types, we may find that the inference *is* formal after all.

We can see that this coloured ball inference may easily be captured in type theory by starting from a relation between say *Object* and *Colour*, expressed in terms of the proposition that '*x* is partly coloured *y*', which depends on the types of objects and colours. If we prefer, we can convert this to a map, *Colour of*, from objects to sets of colours, subtypes of *Colour*. Then our asserting that the ball is red means that *red* : *Colour of* (*this ball*). We now define *IsColoured*(*x*) for *x* : *Object* as a proposition depending on objects, where *IsColoured*(*x*) is equivalent to the type *Colour of*(*x*) being inhabited. This involves a construction in HoTT known as 'bracketing' which takes a type, *X*, and yields a type, $||X||$, for which any elements are taken as identical. A type may be empty, in which case its bracket type is empty, but if it is not empty, the associated bracket type possesses a singleton element (UFP 2014, §3.7). Our definition is

$$IsColoured(x) :\equiv ||Colour\ of(x)||.$$

So from *red* : *Colour of*(*this ball*), the rules defining bracket types give us $|red|$ : $||Colour\ of(this\ ball)||$, which is to say $|red|$ : *IsColoured*(*this ball*). In other words, 'This ball is coloured' is true.

The familiarity we will gain with our type theory should make things clearer as we proceed, but let me here deal with an objection that may well have just occurred to the reader. Since Sellars and Brandom never suggested that such an inference could *not* be made formal by the addition of premises, we have achieved little more, if anything, than was possible by making explicit hidden premises (such as 'Anything that is red is coloured') with the resources of predicate logic. But, contrary to this way of thinking, I am taking the emphasis that type theory places on the formation of types, which include all propositions, as pointing us to see that a great deal needs to be in place already before a proposition, or other type, can make sense to us, and so be deployed. My claim is that plenty of common inferences which have made philosophers suspicious that we could possibly be employing formal inference steps to carry them out, since rendering them in ordinary predicate logic would require there to be admitted explicit hidden premises, come to be seen as merely following on from the rules of our type theory, and so are perfectly well formalized by it. The truth of 'Anything that is red is coloured' is a consequence of the presuppositions (the contents of what we will call the *context* in Chap. 2) necessary for the propositions in the supposed material inference to make sense in the first place. And the construction of the predicate *coloured* is not a one-off, but is rather an instance of a general contruction which will produce, say, *occupied* for a house that is lived in, *parent* for a person that has had a child, *used* for a book that has been read, and so on, where the number of instances in each case is ignored. With regard to the opposite of the first of these predicates, González de Prado Salas et al. (2017) present the inference from 'This house is empty' to 'Jane is not in the house' as another case of material inference. But what may be construed as a *formal* account of this inference works by way of a judgement that the type of people living in this house, *A*, is *not inhabited* in the technical sense that there is a map from *A* to the *empty type*, in other words that there is an element of what is denoted ¬*A*.[11] Then given *Jane* declared as a person, we can derive that she does not live in the house.

---

[11] See UFP 2014, p. 47.

Computer scientists speak of *sugaring* to refer to the modification of formal notation into a form which is more readable for humans, giving the impression of *sweetening* the syntax for our consumption. Syntactic sugar does not add to the functionality or expressivity of the language. Similarly, we may understand many expressions in natural language as resulting from the sugaring of type-theoretic constructions. Once *desugared*, we should find that many cases of apparently *material* inference are in fact *formally* derivable.

I believe that there are grounds to hope that large portions of Brandom's program can be illuminated by type theory. Since Brandom's inferentialism derives in part from the constructive, proof-theoretic outlook of Gentzen, Prawitz and Dummett, this might not be thought to be such a bold claim. However, his emphasis is generally on *material* inference, only some aspects of which he takes to be treatable as *formal* inference. For instance, the argument he gives in *Making it Explicit* (Brandom 1994, Chap. 6) for why we have substitutable singular terms and directed inferences between predicates, is presented with a minimal formal treatment, and yet this phenomenon makes very good sense in the context of HoTT when understood as arising from the different properties of terms and types. Indeed, the kind of category whose internal structure our type theory describes, an $(\infty, 1)$-topos, presents both aspects—the '1' corresponding to unidirected inference between types (morphisms between objects) and the '$\infty$' referring to the reversible substitutability of terms (2-morphisms and higher between morphisms). On the other hand, his frequent use of formalisms, even very briefly to category theory (for example, Brandom 2010, p. 14), tells us that he recognizes the organizing power of logical languages. Elsewhere (Brandom 2015, p. 36), he notes that while both Wittgenstein and Sellars emphasized that much of our language is deployed otherwise than for empirical description, whereas Wittgenstein addressed this excess as an assortment, Sellars viewed much of it more systematically as 'broadly metalinguistic locutions'. These additional functions of language pertain to the very framework that allows for our descriptive practices, and are what Brandom looks to *make explicit* in much of his work. My broader suggestion in this book is that we look to locate these locutions in features of our type theory, especially those *modalities* we will add in Chap. 4.

In the epilogue to his *Studies in the Way of Words*, Paul Grice (1989, p. 372) situates himself with respect to what he calls the 'Modernists' and the 'Neo-Traditionalists'. In rough terms, the Modernist holds that first-order predicate logic with identity (possibly with the addition of modalities) should count as 'Logic', and that this Logic is adequate to represent what is worth extracting of the inferential practices of common speech. The Neo-Traditionalists demur, claiming that we wield ordinary language more subtly than anything that can be treated by Logic. Grice goes on to find a middle way, allowing some modifications of predicate logic to capture our inference. So now we find that the introduction of a *new* Logic opens up the space for finer discriminations. One might be an Ultra-Modernist and take modal HoTT to capture common inference, or a Traditional Modernist, and continue to have faith in first-order logic. One may also continue to believe that there are examples of important implicature lying beyond possible treatment by either of these formal systems.

Certainly, it is worth exploring whether modal HoTT can manage the concerns of those who, like Grice, feel drawn to the neatness provided by a logical calculus, yet can see its failings. Sometimes this perceived tension leads people to introduce modifications to the calculus. So Grice allows a form of negation to apply to 'The present King of France is bald' to form 'It is not the case that the present King of France is bald', or 'The present King of France is not bald'. In doing so, he is disagreeing with Peter Strawson (1950, 1964) for whom there is a 'truth gap', these sentences not being true or false, but rather meaningless. Grice notes that with such a negation, of the three components of Russell's analysis – existence, uniqueness, baldness – we would expect the third to be in question here, the existence and uniqueness being assumed. So he is proposing a modified treatment in predicate logic. I provide my own HoTT resolution of this problem arising from a non-referring definite description in Chap. 3. Satisfying disquiet with the predicate calculus without the need for ad hoc tinkering should count as a triumph.[12]

## 1.4 Quadrature

Somewhat along the lines of computational trinitarianism and the Rosetta stone, I am proposing a structural *quadrature*, where common ground is sought between mathematics, physics, natural language and metaphysics, mediated through the new logic of modal HoTT. In doing so, I am following the advice of the nineteenth-century American philosopher, Charles Peirce. In the first of his Harvard lectures of 1898, Peirce provides an architectonic of disciplines. Mathematics is there at the beginning, the study of the necessary consequences of any assumptions. Philosophy's two branches of Logic and Metaphysics are to follow. He describes his position in terms of claims:

> that metaphysics must draw its principles from logic, and that logic must draw its principles neither from any theory of cognition nor from any other philosophical position but from mathematics. (Peirce 1992, p. 123)

Peirce then descends through the lower ranks of kinds of sciences—nomological, classificatory, descriptive—but observes that the general historical trend is upwards: descriptive sciences turn into classificatory sciences, and these in turn become nomological. What, then, of metaphysics and logic?

> Metaphysics in its turn is gradually and surely taking on the character of a logic. And finally logic seems destined to become more and more converted into mathematics. (Peirce 1992, p. 129)

If this seems notably prescient with all that mathematical logic has offered over the intervening years, I wonder whether Peirce might not have been a little disappointed by the gulf that still exists between *philosophical* logic and *mathematical* logic. Consider one of the places where logic has had most to say to metaphysics, namely, the use of modal logic and

---

[12] Using monads, a topic we cover in Chap. 4, Giorgolo and Asudeh (2012) treat aspects of Grice's work on conventional implicatures.

its semantics to theorize the metaphysics of necessity. Philosophers quickly took on board Kripke's possible world semantics, but how many have realized that they are dealing there with a simple case of what mathematicians call a *presheaf over a partially ordered set*? It is the mathematical logicians and computer scientists, and particularly those using category-theoretic tools, that have exploited this to the full. See, for instance, the very detailed summary of category-theoretic successes in modal logic by Kishida (2017).

Peirce would have been delighted to find a construction originating in topology being used in logic, since he was an early admirer of the former subject, which he saw as under-pinning his metaphysical theory of *synechism*. Later in the Harvard lecture just mentioned, he invokes the example of the formation of topology as an illustration of how unfounded are the fears that there is little to hold mathematics together as a discipline, since it is merely the field of necessary consequences of freely made assumptions. He notes that through very different paths and in complete ignorance of one another, the two mathematicians Listing and Riemann had been led to make the same assumptions. This idea that something much more profound than expressibility within set theory holds mathematics together is still very current. It seems that there are only so many natural assumptions to make.

Now, in the years since Peirce, plenty has been said meta-philosophically about the way to conduct metaphysics. Should we follow our best science, should we describe our common ways of thinking about the world, or should we employ specifically philosophical means to resolve the tensions and even contradictions in our use of everyday concepts by improving on them? A useful classification of the kinds of notion deployed in these various approaches is given by Casati and Varzi (2008) to introduce their own metaphysical work on the nature of *events*:

- a pre-theoretical, common-sense (CS) notion;
- a philosophically refined (PR) notion, where the refinement is dictated by endogenous a priori considerations—e.g., considerations about certain internal inconsistencies of the CS-notion;
- a scientifically refined (SR) notion, where the refinement is dictated by exoge-nous empirical considerations—e.g., considerations about the explanatory value of event-like notions for theories of space-time;
- a psychological notion: the I-representation ('I' for 'internal') of the CS notion, or more generally the I-representation that subserves the explanation of a number of cognitive performances. (Casati and Varzi 2008, p. 33)

Philosophers certainly approach their subject matter employing each of the four varieties of notion, and in so doing some consider themselves to be doing metaphysics. For instance, Casati and Varzi explain how Strawson's 'descriptive metaphysics' aims 'to spell out the content of our prereflective thought or perception of the world, hence, in our terms, the structure of CS-representations'. (ibid., p. 48).

With *philosophical refinement* we see a move beyond the description of common sense: 'Descriptive metaphysics is content to describe the actual structure of our thought about the world; revisionary metaphysics is concerned to produce a better structure'

(Strawson 1959, p. 9). Casati and Varzi present a view of the philosophical refiner as someone carrying out such revisionary work, someone who might look, for example, to refine the notion of identity so as to resolve the *Ship of Theseus* puzzle or to refine our notion of vague predicates to resolve the *Sorites* paradox. Ordinary language philosophers, even if not wishing to engage in metaphysics at all, share with Strawson a suspicion of such work.[13] Here, where there is no apparent fact of the matter as far as our common-sense judgement goes, why not say that the indeterminacy just isn't to be settled? When such an issue does matter, as in the naming of a ship and the associated litigation as to ownership of the vessel, or as to the right to the use of a name, the legal system can be brought in to decide. Commonly, good arguments may be made for either side, but what is needed is a decision as to 'how to go on', and in a legal system relying on precedent, such as England's, legal judgement will determine how to go on. Philosophy may speak of this 'going on', but it's not philosophy's job to prescribe how to do so.

*Scientific refiners*, on the other hand, look to the natural sciences to provide the best account of concepts. A recent example of this approach is Ladyman and Ross (2007). Here philosophical refiners are criticised for relying on a thoroughly out-of-date scientific background which informs the intuitions they rely upon in their armchair speculation. If we want to discuss *events*, then our first port of call should be to see what quantum mechanics and general relativity have to say.

Finally, there is the question of the reliance we should place on the findings of psychologists and linguists who can tell us about ordinary linguistic cognitive performance or about our everyday perceptual abilities. How do we speak about events? How do we individuate them visually? Whether we must speak of *representations*, as in the 'I-representation' underlying a notion, is a moot point. But we should expect to reach beyond what is accessible from our armchair to the results of experiments in vision or language processing, to cross-linguistic analysis, and so on.

To articulate my quadrature proposal, then, I need to situate it in relation to these seemingly disparate approaches to the study of philosophical concepts. First off, I certainly have an antipathy towards a pure, armchair philosophical refinement which looks to regiment our ordinary concepts with predicate logic. But one can depart from such a position in a number of ways—by leaving the armchair and talking to linguists and psychologists or mathematicians and natural scientists, or by using a different logic. What I want to explore now is how we can come to see these departures as less disparate than they have come to appear.

The *philosophical refiners* often look to consider concepts in extreme circumstances, where they believe their refined formulations should be able to survive. For example, when questioning the nature of identity, I may summon to my mind a space containing only two perfectly similar spheres (Black 1952). I need my account of identity in general to tally with my intuitions on identity in this case. If my account implies that these spheres do not possess separate identities, and this goes against a firm intuition that there are two different balls with their own identities, then I had better change my account.[14] This is somewhat similar to the

---

[13] See Hacker (2001) for a Wittgensteinian critique of Strawson's *rehabilitation* of metaphysics.
[14] A treatment in HoTT of this case of identical balls is given in §3.2.2.

way mathematical concepts are tested. If I have defined a property of a class of spaces, such as *connectedness*, I need to know how it applies to extreme cases, such as the empty space. A definition in terms of there being precisely one connected component to a connected space rules out the connectedness of the empty space, while a definition in terms of there being no non-constant map to the discrete two-point space rules it in. Formally defining a concept makes the range of application clear. It is not surprising, then, to see a ready reliance by philosophical refiners on the calculus believed to pull the strings in mathematics, namely, predicate logic and variants.

However, with predicate logic turning out to have less than was hoped to say about either mathematics or natural language, we might have thought the Fregean–Russellian experiment to treat them on the same basis to be over. There may now seem to be no particular reason to expect that a formalism suited to mathematics and physics should have a great deal to say about natural language. Then, if sufficient distance exists between these symbolic forms, metaphysics can at best 'split the difference', as it were. Our metaphysics might follow physics and its mathematical formulations, as the scientific refiners would wish, or it might remain more closely associated with natural language and cognition as governing our everyday conceptualization of the world, along the lines of a 'descriptive metaphysics' or the psychological approach. Alternatively, it might set out on its own path, as the philosophical refiners wish, arriving at concepts with marked differences to our common-sense and scientific concepts. Those urging the latter option, such as Barry Smith, a harsh critic of the use of predicate logic in philosophy,[15] may seek to generate their own formal system to develop their metaphysics, as in *Building Ontologies with Basic Formal Ontology* (Arp et al. 2015).

But don't we have reason to hope that the gap under discussion between mathematics and natural language might be narrower than believed? After all, we have evolved to survive and flourish in a world which we know is describable mathematically. So our orientation to the world should reflect an 'effective' physics. It would not be surprising, then, that our cognitive processes and the framework of our language bear within them the seeds of full-blown mathematical ideas. Perhaps we might appeal to Ernst Cassirer for the start of such an account:

> For all the concepts of theoretical knowledge constitute merely an upper stratum of logic which is founded upon a lower stratum, that of the logic of language. (Cassirer 1925, p. 28)

Cassirer continues:

> All theoretical cognition takes its departure from a world already preformed by language; the scientist, the historian, even the philosopher, lives with his objects only as language presents them to him. (Ibid., p. 28)

Even when theoretical cognition 'takes flight' from its departure point, just as Cassirer believes modern geometry soars from the starting point provided by the structure of

---

[15] See his 'Against Fantology' (Smith 2005).

ordinary vision, as discussed in Chap. 5, we can see the mark of its origins retained within it. Similarly, the physicist's gauge field can be seen as arising by taking flight from a basic type theoretic construction suitable for depicting the '*and*' of natural language, known as the *dependent sum*.

We can pursue this closer association of the structure of natural language with that of theoretical cognition without submitting to the philosophical tendency to want to neaten them too far. An excellent antidote to this tendency is Mark Wilson's *Wandering Significance* (2006). Drawing lessons from the mathematical physics of material science of the nineteenth century, we learn of the work required to maintain the image provided by the tidy classical picture. A term such as 'hardness' provides him with an excellent illustration of the meandering ways the term took as scientists found how different senses arose from diverse investigations within the material sciences. Here, as with the philosophical refiners, concepts are put under pressure, but unlike with the former, there is empirical feedback from the world. A patchwork of uses results.

Still, an inclination to side with Wilson, and, for somewhat related reasons, with ordinary language philosophy, goes unchecked due to the false appearance of a lack of success on the part of formalizers. While my primary motivation in learning about HoTT has been to understand the light it sheds first on mathematics and second on physics, I have been pleasantly surprised to find a wide range of work in natural language semantics employing the dependent type theory at its core. However, the vast majority of this work takes place in disciplinary subjects other than philosophy. In these settings, to use untyped predicate logic would place you in a tiny minority. The lambda calculus or a simple Montague-style type system is more the norm, but increasingly work on dependent type theory is occurring.

Something you notice on entering these fields is that there are plenty of people working in logic, computer science, theoretical linguistics who are reading and referring to philosophers such as Frege, Reichenbach, Kripke, Strawson, Geach, Grice, Lewis, Davidson and Vendler. Zeno Vendler is today perhaps the least well-known of these philosophers, but to those linguists who work on event structure he is a household name. While a committed Wittgensteinian, he also believed that there was a systematic story to draw from the regular transformations seen in our grammar:

> What the Oxford philosophers did in an informal manner, what Austin tried to develop into a 'linguistic phenomenology', can be pursued and made more cogent and powerful by drawing upon the resources of contemporary linguistics. (Vendler 1965, p. 604)

Thus, he strikes a happy medium between two extremes: formalism with little sensitivity to the real structure of natural language on the one hand, and certain strands of ordinary language philosophy hostile to any regimentation, on the other. With his advice to look more closely at what the linguists were finding, Vendler was suggesting that philosophy make a step similar to that made from the amateur scientist stage of English Victorian natural science, where gentlemen of leisure might pursue botanical specimen collection, to a professionalized form. This is a call similar to Ladyman and Ross's advice to leave the armchair, but one which asks us to follow Casati and Varzi's fourth 'psychological' route.

What happens to fields where an original philosophical activity has given rise to a body of research conducted outside of the subject? How ought the philosophical heirs of these philosophers deal with the heritage of their ancestors in these other fields, especially when at least at face value they are engaged in similar kinds of activity? Well, it's not as though all of these ancestors are clearly identifiable straightforwardly as philosophers in the first place. Frege worked in a mathematics department, and several have thought it proper to understand him that way (see, for example, Tappenden 1995). Perhaps we should be careful about policing disciplinary lines. There are certainly people in computer science departments who are philosophically trained and work on matters closely related to philosophical ones, as I have discovered from arranging workshops on the interface between type theory in computer science and in philosophy. We often find that linguists take up the work of philosophers, and later computer scientists follow suit. For instance, *graded modalities* appeared first in the early 1970s in the work of philosophers such as Goble (1970), Fine (1972) and Lewis (1973). The topic was developed through the 1980s in linguistics, such as by Kratzer, where it is still an actively researched topic (see Lassiter 2017). Meanwhile, computer scientists consider a range of modalities in their type theories to deal with a host of topics such as security, resource usage, certainty, and so on (see Gaboardi et al. 2016).

What of other experiences of philosophy engaging well with its offspring once they've grown up? The obvious candidate here would be physics. The philosopher of physics today will likely have a good grasp of at least relativity theory and quantum mechanics. They clearly don't mind that quantum mechanics was produced by non-philosophers. These days they may also be versed in attempts to construct a quantum gravity, perhaps string theory. Shouldn't then a philosopher of language know what a computational linguist has to say? Won't to fail to do so be as though a philosopher of physics only treated work he or she thought of as deriving from philosophy, say, Descartes's 'Nature abhors a vacuum', or Leibniz on the contradictions in accepting an absolute space.

Perhaps the results of linguistics (formal semantics and pragmatics) aren't sufficiently well established, unlike in the case of the physicists. Or maybe *our* questions about language are different ones, though again we should explain why the reference on the part of linguists to so many philosophers, such as Reichenbach on tense and Vendler on events. Here I will be siding with those who are looking to maintain relations, as are the authors of a handbook article on modal logic and philosophy:

> Modal logic is one of philosophy's many children. As a mature adult it has moved out of the parental home and is nowadays straying far from its parent. But the ties are still there: philosophy is important for modal logic, modal logic is important for philosophy. (Lindström and Segerberg 2007, p. 1153)

An important topic for later in this book, modal logic has certainly lived quite a life away from home, especially in computer science departments.

Returning to natural language, it may turn out that there is an untameable motley of linguistic phenomena. But we could at least acknowledge that there are patterns worth exploring. If, as we will see in Chap. 2, Davidson could insist that events form a sort, different from the sort of objects, it is linguists following Vendler who have given us our best account of the structure of events. For example, as we will see, we now know why we do not say

- Paul broke the plates off the table,

while we happily say

- Anna brushed the crumbs off the table.

And somehow we learn how to parse a sentence as complicated as:

- It took me two days to learn to play the Minute Waltz in sixty seconds for more than an hour.

Surely this must involve us having the ability to frame descriptions through intricate event structures. Perhaps no formal tools will help, but we should at least give our best ones a try, and those, such as the type theory I consider, which are battle-tested in computer science, mathematics and physics should have a better chance.

To the grammaticality of sentences, we can also add their implicatures as sources of hidden structure. If these latter arise formally, then I would need to show the type-theoretic rules allowing such conclusions to be drawn from the *Minute Waltz* proposition as:

> I presumably started by not being able to play the Minute Waltz sixty times in a row each within a minute. I didn't achieve this feat within the first day. I probably improved from beginning to end. At the end of the two days I can play the Minute Waltz at least sixty times in a row.

I shall take it that an excellent way to gain insight into the structure of the conceptual network that constitutes one's ontology is success in representing such implicatures. So, our working hypotheses as we move forward, then, are as follows:

1. Ordinary language, designed to deal with the complex array of natural, psychological and social phenomena encountered in our daily lives, possesses rich but undeveloped indications of structure.

2. To locate these indications of structure we need to focus in particular on well-formedness, and on presuppositions and implicatures.

3. The starting points of deep mathematical concepts are to be found in everyday cognition, underpinning our perceptual and linguistic capabilities.

4. Some of this structure receives valuable development in modal HoTT, which can then be employed to extend the capacities of ordinary language. The enhancements are especially relevant for mathematical physics.

## 1.5 Conclusion

Homotopy type theory is new on the scene, even if years in the making, and its *modal* variant has barely had a chance to see the light of day. All the same, opportunities for philosophers are littered about like glistening nuggets on a freshly discovered gold field. The task I set myself in this book is to write enough to encourage other pioneers to remake analytic philosophy by joining the search for application of modal HoTT in any or all of:

1. Philosophy of language
2. Metaphysics
3. Philosophy of mathematics

We will also touch on a fourth area:

4. Philosophy of physics

leaving a fuller treatment to the future.[16]

The target of this work as a whole is **modal homotopy dependent type theory**. Behind each part of the expression is a bewildering body of thought, including much that is generally unfamiliar to philosophy. The working lives of hundreds of people have been invested in its construction. We will gently peel its layers over successive chapters in reverse order:

- Type
- Dependent
- Homotopy
- Modal

So now in Chap. 2 we take our first two steps, and discover why we should think first in terms of *types*, and second in terms of *dependent* types. In Chap. 3, we see how only the lower rungs of a hierarchy of forms of identity have been the standard fare of philosophy. The final *modal* step will be taken in two parts, chapters 4 and 5. The former will introduce modalities similar to those with which philosophers are already well acquainted in modal and temporal logic. The latter will discuss unfamiliar modalities, in particular the *cohesive* ones, which have been devised to represent geometric constructions.

I should like to point out here that I am very far from wanting to making the claim that modal HoTT contains the only aspects of category theory from which philosophy can profit. There's a whole world of ideas out there waiting to be studied, from the use of monoidal categories in the analysis of meaning in natural language (see, for example, Coecke, Sadrzadeh and Clark 2010), to network theory and stochastic mechanics (see, for example, Baez and Biamonte 2018). However, I do maintain that it is a particularly rich place to look.

As a general piece of advice, if a philosopher comes along with a new formalism to sell, making promises that it can cure all ills that avail you, you should be very suspicious. Suspicious that what may just be snake oil can really cure your ills; suspicious that you

---

[16] But see Schreiber (2013), (2014a) and (*forthcoming*).

personally really have these ills; and suspicious that your conditions are indeed ills at all. With this warning in mind, here I come telling you that some ailments in our discipline arise from its use of inappropriate tools, in particular our reliance on untyped predicate calculus, a result of not engaging with those outside of philosophy who are developing and using more appropriate tools. I can only beg the reader's indulgence that they give modal homotopy type theory a fair hearing, and hope they will be persuaded by what follows.

# 2

# Dependent Types

## 2.1 The Need for Types

*Once upon a time, there was a kingdom where everyone was sad. The townspeople were sad. The farmers were sad. Even the King and Queen were sad.*

From the standard untyped perspective, were I to seek to render the opening sentence in formal terms, I might first rewrite it as:

- There is a time and there is a place such that everyone in the place at that time is sad.

I can then represent this sentence as something like

$$\exists x \exists y \forall z (Time(x) \ \& \ Place(y) \ \& \ Person(z) \ \& \ InAt(z, y, x) \to Unhappy(z, x))$$

for a predicate of timed location *InAt*. The question then arises as to what it is I am supposed to be quantifying over in this proposition. It would appear to be a domain whose elements include at least all times, places and individuals. What else should be thrown into the pot to make a coherent domain out of this disparate collection of items? While some philosophers, notably Timothy Williamson (2003), advocate unrestricted quantification over 'everything', merely possible entities included, other researchers in philosophical and mathematical logic have sought highly restrictive forms.

For instance, the type theorist of a certain stripe will take there to have been preparatory work already completed before the formation of the proposition. We would have declared the existence of some types: *Time, Place, Person*. We may well have distinguished between a type of *Time intervals* and a type of *Time instances*.[1] Type declarations provide criteria for the behaviour of their elements, including, importantly, identity criteria. Then there would perhaps be a function from Person to Time intervals, marking the life of the individual, and a function locating their whereabouts at times in their lives. Further, one could construct a type family of people alive at a certain time and living in a certain place, which varies

[1] We will do precisely this in Chap. 4.

*Modal Homotopy Type Theory: The Prospect of a New Logic for Philosophy*. David Corfield, Oxford University Press (2020). © David Corfield. DOI: 10.1093/oso/9780198853404.001.0001

with time and place, $Person(t,p)$. The original proposition could then be constructed itself as a type, where its truth would amount to producing an element of that proposition, so here an element of the type of triples of the form $(t,p,f)$, where $t : Time$, $p : Place$ and $f : \prod_{z:Person(t,p)}(Unhappy(z,t))$. This last element, $f$, is the kind of thing which when given a person from time $t$ and place $p$, spits out evidential warrant of their unhappiness then. Notice that each instance of quantification takes place on a specified type, whether a simple type, such as $Time$ or $Place$, or one formed from other types, such as $\prod_{z:Person(t,p)}(Unhappy(z,t))$.

So one very important difference of type theory from first-order logic is that we don't have untyped variables, that is, ones ranging over some unspecified domain. When we consider the classic 'All ravens are black' as $\forall x(Raven(x) \rightarrow Black(x))$, its truth depends on whether it really applies to *everything*. With unrestricted quantification, this really should mean everything you take to exist: a bird, a table, the Queen of England, a number, perhaps even your filial duty. The failure of all non-bird entities to satisfy $Raven(x)$, along with all non-raven birds, means that the implication is classically true in their case, whatever their colour. Of course, we typically smuggle in 'sensible' restrictions with talk of a *domain*, but this is a situation where we ought to avoid such vagueness. Such imprecision troubles us in the so-called *Ravens* paradox where the negations of $Raven(x)$ and $Black(x)$ are thought to make sense, and the supposedly logically equivalent '$\forall x(non\text{-}Black(x) \rightarrow non\text{-}Raven(x))$' is constructed. Once again, we should reflect on what the domain is taken to be. Are the times, places and people needed for the previous proposition included? For the type theorist, on the other hand, $Black(x)$ must be properly typed. It is a proposition depending on an element of some other specified type. How to delimit this specified type is then an important question, whether here a broad type of space-time continuants or a narrow type of birds. In the solution to the *Ravens* paradox, which proposes that an instance of a non-black non-raven *does* act marginally to confirm the universal statement, it is commonly argued that the relative numbers of non-ravens to ravens matters. But, as we shall see, we only have the individualization permitting such counting when we work with a reasonably prescriptive master type over which quantification is taking place.

Why logical formalisms, one of whose aims is to capture the structure of natural language, should have foregone typing is worth pondering. Natural languages are quite evidently marked by typing information to greater or lesser degrees. When I ask a question beginning *Who* or *When*, I expect an answer which is a person or a time, even if I hear 'April'. Some languages are more thoroughly lexically typed than English. In Japanese, when counting objects, aspects of their nature are reflected in the choice of appropriate counter words, *hai* for cups and glasses of drink, spoonsful, cuttlefish, octopuses, crabs; *dai* for cars, bicycles, machines, and so on. Swahili has *eleven* classes of lexically marked nouns. But even without explicit lexical markers for kinds of noun, the depth of type structuring is revealed by subtleties in the grammatical legitimacy of sentences. This is shown clearly by similar pairs of sentences, the second in each of which is problematic:

- I ran a mile in six minutes.
- *I ran a mile for six minutes.
- I had been running a mile for six minutes, when I fell.

- *I had been running a mile in six minutes, when I fell.
- I spilt some coffee on the table, but it's been wiped clean.
- *I have spilt some coffee on the table, but it's been wiped clean.
- She brushed the crumbs off the table.
- *He broke the plate off the table.

As we shall see later, this may be explained by the compositional structure of event types. The ontology of events as distinct from objects will feature significantly throughout this chapter.

Type theory has found many supporters in computer science for a number of reasons, not least that a desirable feature of a program is that it processes data of specific kinds, and calculates answers and performs actions of specific kinds. At a basic level, we would like our machines to tell us that something is amiss when we enter a number on being asked our name. But we would also like the internal operations of the program to respect the types of the data. Type-checking, either while being compiled before the program is run ('static') or while the program is running ('dynamic'), provides assurance of the correctness of the procedure, as expressed by the slogan 'well-typed programs cannot go wrong'.

Typing usefully wards off category mistakes in natural language and is often supported by syntax. However, plenty of jokes play on the reverse phenomenon, arising from ambiguities of different type readings. A collection of words which may be parsed in two ways when spoken, and sometimes also when written, can give rise to very different readings. So when the joke sets off with *A termite with a toothache walks into a pub and asks, 'Where's the bar tender?'*, we are typically hearing a request for an individual's location, as if a reasonable answer is *'The bartender is in the cellar.'* When we discover the joke is supposedly complete from the expectant look on the comedian's face, we rapidly engage in some re-parsing to take into account the inclusion of the detail of the termite and its ailment, and duly realize that the sort of answer the question requires is *'The bar is tender over by the wall.'*

Type theories come in many shapes and sizes, one important axis of variation being the richness of the type structure. On one end of this axis we find the kind of type theory used by Richard Montague (1974). As developed from Alonzo Church's work, Montague worked with a sparse type theory with basic types of individuals, $e$, and truth values, $t$. From these one can build up compound types, such as properties, of type $(e \rightarrow t)$, i.e., assignments of 'True' to individuals with that property. This then allows quantifiers to be of type $(e \rightarrow t) \rightarrow t$, taking properties to truth values, and so on, such as when a property holds for at least one individual.

But if *the Battle of Hastings* and *Vlad the Impaler* are two terms of type $e$, it seems as though we will be able to ask of them whether or not they are identical. We are marginally better off here than when working in a completely untyped setting, for example, by not being able to ask whether the entity *Julius Caesar* is equal to the property *crossed the Rubicon*. However, there is a venerable philosophical tradition, associated in recent decades with the name of Donald Davidson, which argues for a fundamental distinction in our ontology between *objects* and *events*.

Davidson was led to take *events* as a basic ontological category for many reasons, not least because we can individuate them and yet apply a range of predicates to them which differ from those suited to objects. Indeed, events are plausibly not objects. We don't ask of an event what its colour is, but where it took place, its duration, and so on. We don't ask of an object whether it has occurred or has ended, whether it was expected, and so on. A type theory which allows for several basic types seems desirable, then. The type theory we consider in this book will allow for many basic types, along with those derived from them. This will raise the question of how many basic types there are. *Vlad the Impaler* is apparently an element of the kind of rulers of Wallachia, the kind of political leaders, the kind of humans, the kind of primates, the kind of mammals, the kind of living things, the kind of objects. The Battle of Hastings is an element of the kind of mediaeval battles on English soil, the kind of battles, the kind of events, and so on. It might seem that subtypes are being formed by increasing specification. The type *Human* will be something like *Primate of a certain sort*, primates will be mammals of a certain sort, mammals will be animals of a certain sort, and so on, right up to objects of a certain sort. If we carve out more specific types by some kind of operation on the general types *Object* and *Event*, why not take the further step and unite these types, and maybe others, in a supertype of individuals, as Montague does? Then objects are individuals which are object-like and events are individuals which are event-like. Let us see why we might want to resist this move.

Psychologically, at least, we seem not to conflate across this boundary. One useful way to test this is to see when we can apply 'or' to them. Gazing on a distant shoreline, I might wonder whether I'm seeing a seal or a smooth rock. Someone looking on in the Senate asks 'Is Brutus greeting Caesar, or is he stabbing him?' These latter are events, and indeed events taking place at a certain time and place and involving the same protagonists. On the other hand, we don't say 'that is either a chair or a battle', but looking at a rowdy group of youths in the distance we say 'it's either a street party or a fight'. Widening the range of types to include colours and places appears to make things worse. Imagine saying 'That's either the colour red or it's a marriage or it's the corner of my bedroom.'

When someone simulates a fire with a piece of red cloth being billowed about from below and I ask 'Is it a piece of red cloth or a fire burning?', it might seem that I'm crossing the object/event divide, but then to the extent that a fire is considered an event by being a process, it could be better to construe my question as asking which event type is occurring: 'Is that a piece of red cloth billowing or a fire burning?'

As Casati and Varzi (2015) explain, some have suggested that an object is a degenerate process, 'a thing is a monotonous event; an event is an unstable thing' (Goodman 1951, p. 286), or more generally for a spectrum between the 'firm and internally coherent' object (Quine 1986, p. 30) and the rapidly unfolding event. It has also been suggested that events have a more definite time demarcation, but may have loose spatial boundaries, while objects tend to have better-defined spatial boundaries. Matters are made more complicated by entities whose constituents change, preserving to some extent shape or place. Animals maintain a certain body shape while replacing their constitutive elements slowly, yet may move place rapidly. Rivers replace their constitutive elements rapidly, yet are more static than animals. Fires can vary from the steady flame of a candle to the raging forest fire.

One revolutionary approach is looking to replace talk of objects by *processes*, at least in the biological realm:

the world—at least insofar as living beings are concerned—is made up not of substantial particles or things, as philosophers have overwhelmingly supposed, but of processes. It is dynamic through and through. (Nicholson and Dupré 2018, p. 3)

Now, rather than take the stability of living entities for granted, the focus is on how such stationarity is produced in the flux. This requires a radical revision to the way we think and indeed speak. However, since I will be drawing heavily on natural language use throughout this chapter, I will follow those who argue for the equal priority of occurrents and continuants.

Another approach to type classification comes from individuation and counting. We will come to see that our type theory contains some types which are *sets*. This is meant in the technical sense that for any two elements of such a type, their identity forms a proposition, again in a technical sense. This in turn means that there can be no multiplicity of how two elements are the same, merely a question of whether or not they are the same. So a set is a type such that it is only a matter of *whether* its elements are the same, not *how*. This will be expressed formally later, but let us note here that in everyday life a kind of thing generally allows some form of individuation. Typically, the more specific the kind, the clearer this becomes. Many of the kinds of types we employ in ordinary language behave as sets in the technical sense. As might be expected, due to this property of individuation, any set has a cardinality. Relative to a specified type which is a set, individuation should be such as to allow counting to take place. Even if this is difficult, as with the types *Hair on my head* and *Leaf on that tree*, it makes sense to imagine that there is a total number. On the other hand, other types, and typically less specific ones, do not allow easy individuation. Until one specifies a variety of *object*, it is generally the case that counting is problematic. You don't ask, looking in on a kitchen, how many objects there are in there, but to ask for the number of plates is fine.

It has been proposed (see, for example, Wellwood, Hespos and Rips 2017) that there is an important analogy between objects and events in terms of their relationships with their respective constituents:

object : substance :: event : process.

In the case of substances, we tend to identify portions as constituting objects and apply plural noun forms when we see them as reasonably regularly shaped. A football on a soccer pitch possesses this feature very clearly, a sandcastle on a beach somewhat less clearly, while a puddle on a pavement may be very hard to identify. Similarly, to consider a process as constituting an event requires some order, in particular a temporal boundary. Sharp boundaries such as provided by a football match make for easy identification and counting.

Discriminating and counting *events* thus requires us to carve out discrete parcels of processes of a kind. Gazing over an Olympic gymnasium, we may ask how many events are occurring now: the 2012 Summer Olympics, a gymnastics competition, women's gymnastics, the vault, an athlete is running, a step is taken, and so on. Again, to be asked how many events occurred in the Second World War makes little sense, but one could imagine being asked to estimate the number of a better specified type, such as rifle shots on Juno beach during the first day of the Normandy landings. We now obsessively monitor the steps we take in a day, even though for any particular foot movement its status as a step may not be clear, such as when you make a small adjustment of stance as someone brushes past you. Similarly, we make a reasonable attempt to individuate hurricanes and tropical storms, and even name them alphabetically, all the time waiting for them to 'lose their identity'. We will return to the details of the structuring of events after we have understood ways to express their composite nature.

As for entities on the cusp between events and objects, in a hot and dry summer there may be a number of forest fires in a locality. One threatening my house right here may be taken as different from another one raging ten miles away, but whether to count two patches of the same burning hillside as the same fire or distinct ones may not be obvious. One reason for the difficulty here is that there is little internal cohesion to a fire—it spreads, merges and separates all too easily. On the other hand, some kinds of dynamical processes do lend themselves to individuation and counting. For Collier, *cohesion*, understood as 'an equivalence relation that partitions a set of dynamical particulars into unified and distinct entities' (2004, p. 155), is responsible for this.

> Cohesion is an invariant, by definition, but more important, it determines the conditions under which something will resist both externally and internally generated disruptive forces, giving the conditions for stability. (2004, p. 156)

Such cohesion comes in degrees, so that one can take a flock of birds to be *weakly* cohesive. Counting can thus be seen to be no easy matter, and all but impossible if taken to range over some domain which includes objects, events and other kinds. We need, then, to distinguish kinds or sorts, and to do so we need to specify the conditions for membership of these kinds. Taking his departure from Collier's ideas, DiFrisco claims that

> when the interactions of the relevant type meet the cohesion condition, we get access to a dynamical explanation grounding the *sortal* for the individual in question. (2018, p. 85)

This then allows for individuation:

> processes are identical iff they have all of the same cohesive properties, including cohesion profile and cohesion regime, and they occupy the same spatio-temporal region. (Ibid., p. 85)

Let us now consider what other kinds we must contend with.

We must be careful not to rely solely on words to extract the relevant types. A point often noted by linguists is that terms apparently designating single entities of one type may be better understood as falling under different but related ones. This is the phenomenon of *copredication*, such as when we say *Lunch today was long but delicious*. On the one hand, we take lunch here as an eating event, on the other, as a collection of foodstuffs. The word 'book' displays a similar ambiguity. *I have two books on my shelf. Each contains the three books of 'The Lord of the Rings'. How many books are there: two, three or six?* Of course, it depends on what we mean by 'book'—a physical tome, an information source or an instance of an information source. Such ambiguity renders the following sentences strange:

- John has an extensive library of one thousand books. But he only has one book. They're all copies of *War and Peace*.
- I have *War and Peace* as an electronic book. My reader is light, and so is the book.

In other cases, we may use different terms to disambiguate. Rather than ask at an airline's board meeting how many people have been carried this year, it is likely to be of greater interest to know how many trips have been taken, usually phrased as the number of passengers. Of course, it may be important to know as well how many different people travelled on your airline to understand frequency of flying.

The ease with which a restricted kind of middle-sized object allows for counting must surely be due to the underlying solidity and cohesion of its substance. If we have defined the type *Cup* in terms of objects which are hand-sized and used to contain a liquid for consumption, the identity condition for *Cup* can piggyback on that for object. But there is a part-of relation on objects, so a handle is part of a cup and also then an object in its own right. Given two elements of the same specific kind, identification is more straightfoward: how many cups in the cupboard, how many handles on this cup.

Sometimes an object may support more than one instance of another type. The artist Ivon Hitchens painted many semi-abstract landscapes from around his dwelling in Sussex. His choice of colour made it quite feasible to turn his elongated rectangular works upside down to present a different image. Indeed, he joked that his clients were receiving two paintings for the price of one. Here an individual painting is an object plus an intended orientation. If I hang the Mona Lisa the wrong way up, this is a subversive act.

Of course, other types most definitely are not supported as material objects: melodies, manners of walking, colours. George Harrison could be found guilty of using the melody from *He's So Fine* when composing *My Sweet Lord*. The type *Colour* has its own identity criteria in terms perhaps of side-by-side comparison of instances. Two colours are the same if when juxtaposing patches of these colours in good lighting conditions, we cannot tell them apart. We also use colour charts to provide standard reference points. When you say 'I love Chicken Bhuna', you are referring to an element of *Dish*, which generates the expectation that all instances of *Plateful of Chicken Bhuna* should be liked by you. What kind of thing is a dish? It doesn't have an associated place, yet it has a flavour.

Places rely on the fixity of environments. We can bury a pile of treasure at a spot and return to find it decades later. We can also have a place on a boat to keep track of our possessions while circumnavigating the world. Place is not generally a space-time point, but rather a path traced out by some stable configuration of objects. *'To think we were in the same place as each other on New Year's Eve without knowing.'* Individuation occurs here (my trip to the US has me visiting three places: Boston, New York City and the Niagara Falls), but inevitably there will be a spatial structure to this type. Places can be closer or further from each other, and I'm unlikely to say that I visited ten places in France, namely, Paris and nine sites in the city of Bordeaux. The formal introduction of spatial features in our type theory will be deferred until a later chapter.

Alongside evidently spatial types, it seems plausible that many everyday types contain further structure relevant for counting. Presented with a collection of seven oil paints, I may be disappointed as a beginner looking to equip myself with the basics to find that these are: royal blue, cobalt blue, French blue, Prussian blue, ultramarine blue, manganese blue and cerulean blue. The type of colours would seem to have a richer structure than being a mere set. Indeed, perhaps 'blue' and 'red' are better taken as designating a range of colours. It is common also to use spatial descriptions here, so that it is hard to imagine anyone taking Cobalt blue to be 'closer' to burnt umber than to Prussian blue. Similarly, with animals, we are disappointed when visiting a zoo said to hold animals from 91 different species, if these are close relatives. We want to see dramatically different kinds, and not representatives of the 91 species of antelope.[2]

Looking beyond what seem to be basic types, we will also have to consider functions, such as a matching of my knives and forks, an assignment of soldiers to their regiments, a grading of the work of a cohort of students. Once these are admitted, the production of a master type of all individuals would seem to push us towards an intolerable inhomogeneity. We never would ask

> Is that burnt umber or a game of football or an elephant or a preference order of candidates or a declaration of war?

More radically, we are going to look to have *propositions* as types too. Let us now see why we should want to view a proposition as a kind of type on the same footing as a set-like type.

## 2.2 The Analogy between Logic and Arithmetic

One good way to motivate this novelty of our type theory in taking propositions as a kind of type is by showing how similarly propositions and sets behave. When I teach an introductory course on logic to philosophy students, I like to point out to them certain analogies between the truth tables for the logical connectives and some simple arithmetic operations, some of which have been known since the nineteenth century.

If we assign the values 1 to True and 0 to False, then forming the conjunction ('and') of two propositions, the resulting truth value is determined very much as the product of numbers chosen from $\{0, 1\}$:

- Unless both values are 1, the product will be 0.
- Unless both truth values are True, the truth value of the conjunction will be False.

It is natural for students, then, to wonder if the disjunction ('or') of two propositions corresponds to addition. Here things don't appear to work out precisely. It's fine in the three cases where at least one of the propositions is False, for example, 'False or True' is True, just

---

[2] For a mathematical treatment of species diversity measures, see Leinster and Cobbold (2012).

as $0 + 1 = 1$. But in the case of 'True or True', we seem to be dealing with an addition capped at 1, something such as would result in a third glass when the contents of two other full or empty glasses of the same size are emptied into it.

Setting to one side for the moment this deviation, the next connective usually encountered is *implication*. A very good way of conveying to students the meaning of an implication between two propositions, such as $A \rightarrow B$, is as an *inference ticket*, a process taking a guarantee for the truth of $A$ into such a guarantee for $B$. Implications seem to act very much like functions. Now, the standard truth table for implication assigns True to each row except in the case of 'True implies False', corresponding to the lack of a map from a singleton to the empty set. So, once again the truth table is reflected in the Boolean arithmetic of exponentials, $1^1 = 1^0 = 0^0 = 1$, while $0^1 = 0$.[3]

We begin to see that propositions as understood by propositional logic behave very much as sets of size 0 or 1. But now we notice further that the connectives can be put into relation with one another. Denoting the conjunction of $A$ and $B$ by $A\&B$, we may further observe the analogy:

- $(A\&B) \rightarrow C$ is True if and only if $A \rightarrow (B \rightarrow C)$ is True.
- $c^{(a \times b)} = (c^b)^a$.

We may reformulate the former as expressing the equivalence between deducing $C$ from the premises $A$ and $B$, and deducing $B$ implies $C$ from the premise $A$. This equivalence is understood by every child:

- If you tidy your room and play quietly, I'll take you to the park.
- If you tidy your room, then if you play quietly, I'll take you to the park.

So what accounts for this pleasant logic–arithmetic analogy? The idea of a proof of $A \rightarrow B$ as a function and its analogy to the arithmetic construction $b^a$ should bring to mind the fact that the latter may also be interpreted as related to functions. Indeed, $b^a$ counts the number of distinct functions from a set of size $a$ to a set of size $b$, so that, for example, there are nine distinct functions from a set of size 2 to one of size 3.

Now we can see a closer analogy between proposition and sets, first through a parallel between conjunction and product:

- Guarantees of the proposition $A\&B$ can be formed from pairing together a guarantee for $A$ and one for $B$.
- Elements of the product of two sets $A \times B$ can be formed from pairing together an element of $A$ and one of $B$.

---

[3] In continuous settings, $0^0$ is generally considered to be of indeterminate value, but what provides a rationale for the answer 1 in discrete, combinatorial settings is relevant here.

And second, there's a parallel between implication and function space, or exponential object:

- Guarantees of the proposition $A \to B$ are ways of sending guarantees for $A$ to guarantees for $B$.
- Elements of the function space of two sets $B^A$ (or $Fun(A, B)$) are functions mapping elements of $A$ to elements of $B$.

This approach to logic is very much in line with the tradition of Per Martin-Löf (1984), and in philosophy with the ideas of Michael Dummett (1991). What homotopy type theory does is take observations such as these not merely as an analogy, but as the first steps of a series of levels. Propositions and sets are both considered to be simple kinds of *types*. In Martin-Löf-style type theories, given a type and two elements of that type, $a, b : A$, we can form the type of identities between $a$ and $b$, often denoted $Id_A(a, b)$. The innovation of homotopy type theory is that this identity type is not required to be a proposition—there may be a richer type of such identities. But we can see now how the levels bottom out. If you give me two expressions for elements of a given *set*, the answer to the question of whether or not they are the same can only be 'yes' or 'no'. We say that the identity type is a *proposition* or *mere proposition*. On the other hand, in the case of a type which is a proposition, were we to have constructed two elements, in this case proofs, and so know the proposition to be true, these elements could only be equal.

So taking identity types of the terms of a type decreases the type's level, and this may be iterated until we reach a point where the type resembles a singleton point, known as a *contractible* type. For historical reasons connected to the branch of mathematics known as homotopy theory, this level is designated $-2$. Mere propositions are thus of level $-1$, and sets of level 0. The hierarchy then extends upwards.

| | |
|---|---|
| ... | ... |
| 2 | 2-groupoid |
| 1 | groupoid |
| 0 | set |
| $-1$ | mere proposition |
| $-2$ | contractible type |

In Chap. 3, we will give a formal treatment of the hierarchy and explore the consequences of allowing higher-level types, ones for which there may be a *set* of identities between two terms or a type of identities of a yet higher level. Here, for the moment, we may think of an easy example of a type for which this is so. Take the type of finite sets.[4] For any two elements, that is, any two finite sets, an element of their identity type is any map which forms a one-to-one correspondence between the sets. In the case of equally sized sets of cardinality greater than 1, there will be several of these correspondences. But then there is nothing further to say about any two such identities, other than that they are the same or different. The identity

---

[4] This is defined via the type family $Fin(n)$ (see §3.1.2 below and Ex 1.9 in UFP 2014).

type of two finite sets is a set. The type of finite sets, therefore, has level +1, since its identity types are sets, of level 0. And so it goes on. One may have types of infinitely high level.

In Martin-Löf-style type theories, propositions are also types, the types of their warrants. In homotopy type theory a type with at most one element is called a *mere proposition* or simply a *proposition*. This terminological decision reflects the fact that Martin-Löf designed his type theories for mathematical use, and in mathematics the word *proposition* is typically used as a claim which is to be established. The decision of the HoTT people to call these 'mere propositions' comes from the fact that once a claim has been proved in a piece of mathematical writing, details of the proof are then generally forgotten. It's just the fact that the claim has been shown to be true that matters, and different proofs of a proposition are not distinguished.

There are two noteworthy types which are propositions that may be formed simply in our type theory. These propositions account for the 0 and 1 appearing in the logic–arithmetic analogy above. To define the empty type 0, there are no elements to construct, but there is an elimination rule which says that from an element of 0 then for any type, $C$, we can construct an element of $C$. This is the type-theoretic version of *ex falso quidlibet*, with 0 as the proposition *False*. On the other hand, to define the unit type, 1, we specify a single element, $\star$ : 1. 1 plays the role of the proposition *True*.

Whether or not the terminology *mere proposition* lasts in the long term, it leads to an important finding: **There is no need to propose a separate logic for HoTT**. The general type constructions suffice. For instance, almost all type theories have a rule that takes two types and generates their product:

$$A, B : Type \vdash A \times B : Type.$$

Elements of the product type are formed by pairing elements of the separate types. When you give two coordinates for a point in the plane or place on a map, 43 degrees north *and* 13 degrees east, the 'and' there is just like the 'and' which conjoins two propositions. A proof of *A and B* is a proof of *A* and a proof of *B*.

In a similar vein, another type formation rule states that for any two types, there is a type of functions between them:

$$A, B : Type \vdash [A, B] : Type.$$

An element here amounts to a construction taking an element of $A$ to an element of $B$. In the case of mere propositions, an element of the function type is a proof of the implication, a mapping of a warrant for $A$ to a warrant for $B$.

So we see that, from the perspective of HoTT, there is no coincidence in some of the analogies we have just observed, and we shall shortly see that this applies to the others. It is a blindspot of other foundational systems, such as Zermelo–Fraenkel set theory, that conjunction of propositions and the cartesian product of sets and implication and function space are treated so differently. HoTT explains these commonalities as arising from the *same* constructions. And, as we shall see after I have introduced *dependent* types, these commonalities may be greatly expanded. For the moment, though, let's consider why, from the category-theoretic perspective, we should expect product operations and function space objects to appear so very commonly. Taking type theories as the syntactical counterpart of categories, the prevalence of products of types coincides with the prevalence of cartesian products in categories. Since, as I shall now show, products in a category emerge as *adjoints* to a very basic duplication map, their prevalence should not surprise us.

As we should expect from the *computational trinitarianism* thesis discussed in Chap. 1, an account of the appearance of constructions whose instances include logical connectives may be given from a category-theoretic perspective. We owe this account to William Lawvere who in 'Adjointness in Foundations' remarked:

> One of the aims of this paper is to give evidence for the universality of the concept of adjointness, which was first isolated and named in the conceptual sphere of category theory, but which also seems to pervade logic. (Lawvere 1969, p. 283)

A few years later, he explained in his paper, 'Metric Spaces, Generalized Logic and Closed Categories', that we should identify logic with a 'scheme of interlocking adjoints' (Lawvere 1973, p. 142). We will be treating adjunctions in detail in Chap. 4, but for the moment let me remark that they constitute an essential part of category theory. As one of the founders of category theory, Saunders Mac Lane, remarked, 'Adjoint functors arise everywhere' (Mac Lane 1998, p. vii).[5] From a certain perspective, it is quite surprising how little knowledge of adjunctions has penetrated into philosophical consciousness when it has been known for fifty years how integral they are to logic, and all the more so in the case of modal logic, as we shall see later.

We can consider a category, $\mathcal{C}$, to be a collection of objects related by arrows between pairs of them. That is, for any pair of objects, $S$ and $T$, in the category there is a collection of arrows between them, $Hom(S, T)$. Think of these arrows as processes, transformations or inferences running from $S$ to $T$. Then at any object there is an identity process, $1_S$, and any two matching processes, $f : S \to T$ and $g : T \to U$, where the target of the first matches the source of the second, may be composed $g \circ f : S \to U$. Composition of a process with the relevant identity process on either side leaves it unchanged, $f \circ 1_S = f$ and $1_T \circ f = f$. Finally, it doesn't matter in which order we calculate the composition of three or more composable processes, so that for appropriate maps $h \circ (g \circ f) = (h \circ g) \circ f$. A functor between two categories, $F : \mathcal{C} \to \mathcal{D}$, sends identity processes to identity processes, and composites of processes to composites of the images of processes, $F(g \circ f) = F(g) \circ F(f)$.

In the case of a *deductive system*, we have a category with propositions as objects, $P$ and $Q$, and a single arrow in $Hom(P, Q)$ if and only if $P$ entails $Q$. Here composition of arrows represents the transitivity of the entailment relation. A functor between deductive systems we might consider to be a translation from one to the other, where the entailment relation is preserved.

A pair of adjoint *functors*, $F$ and $G$, between two categories, $\mathcal{C}$ and $\mathcal{D}$, is such that for any $A$ in $\mathcal{C}$ and $B$ in $\mathcal{D}$,

$$Hom_{\mathcal{C}}(A, G(B)) \cong Hom_{\mathcal{D}}(F(A), B).$$

---

[5] See also the Notes therein on p. 107.

Here $F$ is the *left* adjoint and $G$ is the *right* adjoint, and one may think of them as providing something approximating an inverse to one another. So now, take a single category, $\mathcal{C}$, and then form the cartesian product of $\mathcal{C}$ with itself. This is a category which has pairs of objects of $\mathcal{C}$ as objects, and pairs of morphisms of $\mathcal{C}$ as morphisms. Clearly there is a *diagonal* map, a *functor*, $\mathcal{C} \to \mathcal{C} \times \mathcal{C}$, which sends any object of $\mathcal{C}$, $X$, to $(X, X)$. We now look to form the right adjoint of this functor. If it exists, this will have to send a pair of objects of $\mathcal{C}$ to a single object. Now, given any three objects, $A, B, C$, in $\mathcal{C}$, we know that

$$Hom_{\mathcal{C} \times \mathcal{C}}((A, A), (B, C)) \cong Hom_{\mathcal{C}}(A, B) \times Hom_{\mathcal{C}}(A, C).$$

So for there to be such an adjunction we need a construction, which we call $B \times C$, such that

$$Hom_{\mathcal{C}}(A, B) \times Hom_{\mathcal{C}}(A, C) \cong Hom_{\mathcal{C}}(A, B \times C).$$

This is precisely what the product construction achieves. Loosely speaking, we could say that as far as logic goes, as soon as one admits the possibility of making more than one assertion, it is natural to arrive at the concept of conjunction. Very similarly, the left adjoint to this diagonal functor corresponds to the sum or *coproduct*,

$$Hom_{\mathcal{C} \times \mathcal{C}}((B, C), (A, A)) \cong Hom_{\mathcal{C}}(B + C, A).$$

Pushing on one more step in the web of interlocking adjoints, if our category $\mathcal{C}$ has such binary products, then for any object, $B$, there is a functor from $\mathcal{C}$ to itself which takes $A$ to $A \times B$. Then for there to be a right adjoint, $G_B$, of *this* functor, it would need to be such that

$$Hom_{\mathcal{C}}(A \times B, C) \cong Hom_{\mathcal{C}}(A, G_B(C)).$$

The image of such a functor we denote $G_B(C) = C^B$. Exponentials thus readily appear in many contexts, and the connective '*implies*' is seen to occur naturally. Again, with the notion of an assertion following from another, it is natural to arrive at the concept of implication.[6]

## 2.3 Dependent Sum and 'and'

Ask philosophical logicians to show you their wares with respect to the meaning of a favoured connective, and they'll often plump for conjunction. 'Or' suffers the ambiguities as to whether it is to be understood as *inclusive* or *exclusive*. Implication suffers worse from the common-sense assumption that '$A$ implies $B$' indicates some relevance between the

---

[6] See Patrick Walsh's paper (Walsh 2017) for more along these lines concerning the relationship between logical connectives and adjunctions, including the appearance of identity types in an adjunction. Technical aside: the *counit* generated by the diagonal-product adjunction is a map in $Hom_{\mathcal{C} \times \mathcal{C}}((A \times B, A \times B), (A, B))$ which represents the paired elimination rules for the product type, and in particular for the conjunction of propositions. Similarly, the *unit* of the diagonal-coproduct adjunction represents the paired introduction rules for the sum type. See §4.3 for more on this.

propositions. But good, reliable 'and' appears to work as it is supposed to. *A&B* is true if and only if *A* is true and *B* is true.

Indeed, when so-called *logical inferentialists* explain how they take the rules governing logical connectives to encapsulate their meaning, they frequently illustrate this with the case of conjunction, writing the *introduction* and *elimination* rules something like this:

$$\frac{P \ Q}{P\&Q} \quad \frac{P\&Q}{P} \quad \frac{P\&Q}{Q}.$$

For Paul Boghossian,

> it's hard to see what else could constitute meaning conjunction by 'and' except being prepared to use it according to some rules and not others (most plausibly, the standard introduction and elimination rules for 'and'). (2011, p. 493)

As a counterexample, Timothy Williamson presented an inference that a three-valued logician would take to be valid, but which runs contrary to conjunction elimination. When Boghossian replied that he does not believe that this case 'presents us with an intelligible counterexample to the analyticity of conjunction elimination' (2011, p. 493), he was professing a commonly held view of the meaning-constitutive power of natural deduction rules.

It was a provocative choice, then, when Bede Rundle devoted his paper on conjunction (Rundle 1983) to the limitations of taking 'and' as the logicians' & or ∧:

> On the one hand, I am inclined to think that the standard philosophical treatment of conjunctions like 'and', 'but' and 'although' has been grossly inadequate, no concern being shown for anything more than a narrow aspect of their use, and no investigation of that use being conducted on the right principles. On the other hand, the preoccupation with truth-conditions which has resulted in this defective approach is one towards which it is easy, and proper, to be sympathetic, at least initially. I shall begin by indicating the considerations which might invite our sympathy, and then call upon the example of 'and' to show how the approach is defective. (1983, p. 386)

Rundle had long campaigned in the ordinary language tradition against formalization:

> it is not uncommon to find a philosopher showing an exaggerated respect for what he describes as the 'logical form' of a class of sentences, seeing in his own artifice a fundamental linguistic or logical pattern when in practice the process of extracting it from actual sentences more often than not results in their distortion. (Rundle 1979, p. 8)

But one might have expected 'and' to be ceded to the opposition.

In his 1983 paper Rundle runs through a number of objections, not the least important of which is that we use '*and*' to connect commands, expressions of wishes and questions, even producing mixtures of these:

- Where do you come from and what do you do?
- Put out the dog and bring in the cat.
- He was decent enough to apologize, and make sure you do too.

He does so to cast doubt on the logicians' reading:

> given that the conjunctive role of 'and' is quite indifferent to mood, we should surely be suspicious of any account of its use that makes reference to truth essential. (1983, p. 388)

Well, a die-hard 'conjunctivist' may suggest that we should take a command to be an instruction to bring about the truth that a state of affairs obtains, at which point it seems reasonable that conjunctions of commands are commands to bring about the truth of the conjunction of propositions that the corresponding states of affairs obtain. But looking further into Rundle's arguments, we see that the main charge he brings to bear is that a crucial semantic aspect of 'and' has been overlooked, one which underpins its conjunctive use:

> if I can show that, as far as it exists, the truth-functional character of 'and' can be accounted for by having regard to more general features of its use, a use that is essentially the same whatever the sentence-types it unites, then the supposed priority [in declarative cases—DC] can surely be considered illusory. (1983, p. 388)

In this section, we consider a central element of these 'more general features of its use', which is that 'and' only conjoins on certain occasions:

- We typically do not use 'and' to connect disparate assertions, still less in the case of the supposedly logically equivalent 'but'.

As we saw with 'is' in Chap. 1, there is no reason to commit ourselves to treating a single word of natural language formally in the same way. It is no different here with 'and.' First, we may use 'and' to judge of two elements that they each belong to a single type:

- Fido and Bella are poodles.
- $\vdash Fido, Bella : Poodle.$

Or two elements may jointly be judged to satisfy a relation:

- Adam and Eve are married.
- $\vdash p : married(Adam, Eve).$
- $\vdash (Adam, Eve, p) : Married\ Couple.$

Frequently, we rely on ellipsis to avoid repetition when conjoining judgements, as with the conjunction:

- Adam is a doctor and Eve is an accountant,

which may be again be represented as the claim that two elements belong to a type:

- (*Adam*, *doctor*), (*Eve*, *accountant*) : *People and their occupation.*

In such a case, we may omit grammatical parts, here the repeated '*is*':

- Adam is a doctor and Eve an accountant.

In the case of repeated surface structure, but not underlying type structure, such usage of ellipsis has a humorous element. Gilbert Ryle gives the example 'She came home in a flood of tears and a sedan-chair' (Ryle 1949, p. 22). We see the beginnings of a witticism here, such as in the description of American GIs serving in Britain in the Second World War as 'overpaid, oversexed and over here'. The grammar is suggesting a common typing through the parsing, which the meaning of the words resists.[7]

Rundle points out that *and* is also used to indicate temporal linkage, so that the following two propositions are not equivalent:

- He used to lie in the sun and play cards.
- He used to lie in the sun and he used to play cards.

It seems here in the first sentence that we mean something like:

- He used to lie in the sun, and at that time play cards.

The cardplaying corresponds to a reasonable concentration of subintervals of sunbathing intervals.

Indeed, it is generally the case that when we conjoin two or more distinct propositions, there is some kind of thematic linkage. Rundle writes

> With conjunctions generally some relation will be implied, and with 'and' it is surely that the word is used to signal, rather than to effect, the relevant linkage or association, for all it could provide in the latter capacity is some form of phonetic bridging. (1983, p. 391)

This makes sense of the observation that the former of the following two sentences is clearly the more natural:

---

[7] Ryle says of other similar typing mistakes, 'A man would be thought to be making a poor joke who said that three things are now rising, namely the tide, hopes and the average age of death. It would be just as good or bad a joke to say that there exist prime numbers and Wednesdays and public opinions and navies; or that there exist both minds and bodies' (Ryle 1949, p. 23).

- Jack fell down and broke his crown, and Jill came tumbling after.
- Jack fell down and is twenty years old, and Jill is holidaying in Thailand.

There are thematic connections right through the first of these sentences—Jack cracks his head because of his fall, and then Jill also falls down after Jack. Rundle claims of this sentence that 'the first "and" gives us a compound predicate based on the possibility of thinking in terms of a single, if complex, event' (1983, p. 389). We will take up this point in a later section. As for the second 'and', he speaks of 'things that Jack and Jill then did' or more generally of 'episodes which occurred at that time'. This would be a case, as above, of two terms belonging to the same type.

Let me offer a slightly different diagnosis. Had it scanned satisfactorily, we might have seen 'Jill came tumbling after *him*' appear in the nursery rhyme to emphasize the linkage. The use of such a pronoun is what grammarians call *anaphora*. When Aarne Ranta (1994) treated such anaphoric reference type-theoretically, he deployed Martin-Löf's *dependent sum* construction, which I will now introduce.

It is common to have a situation where for any element of a given type it is possible to form a type which depends upon that element. In the case of days of the month, each time we consider a month, $m : Month$, we judge that the days of that month, $Days(m)$, form a type. Formulating things this way makes sense for the computer scientist who will want to specify that different types $Day(m)$ as $m$ varies have different sizes. Indeed, given the behaviour of leap years, it may prove useful to produce $Month(y)$, depending on $y : Year$. In the former, simpler case, elements of the dependent sum are *pairs*[8] of month and day in that month, such as (February, 24) and (August, 5). We may like to imagine the days of a month lined up on top of a marker for their respective month. We write $\sum_{m:Month} Day(m)$ for the type of dates. But this construction works just as readily for dependent types which are propositions, such as for any month, $m$, whether it is the case that this is a summer month, $summer(m)$. Now the pairs are constituted by a month and a warrant that that month is in the summer, so (August, warrant that August is a summer month).

In the case of a series of propositions where a later one may depend on a component of an element of an earlier one, the same dependent type structure is in play. For example, *It's raining now, and doing so heavily* can be represented as involving the following judgements:

- $\vdash a : It\ is\ raining$
- $x : It\ is\ raining \vdash Heavily(x) : Prop$
- $\vdash Heavily(a) : Prop$
- $\vdash b : Heavily(a)$
- $\vdash (a,b) : \sum_{x:It\ is\ raining}(Heavily(x))$

where $a$ and $b$ are suitable warrants for their respective propositions. The final judgement expresses the truth of *It's raining now, and doing so heavily*. The statement might arise as a

---

[8] This is why some authors refer to the construction as 'dependent *pair*'.

response to being posed linked questions of the form: P? And if so, Q? So here, *Is it raining? If so, is it raining heavily?* The second question only makes sense on condition of a positive response to the first. Without the element *a* of *It is raining*, there is no proposition *Heavily*(*a*) of which to discover its truth. So, similarly, 'Jill came tumbling after (him)' only makes sense if we already have that Jack has fallen.

This phenomenon also arises when we form a proposition by 'truncating' a type. Think of a questionnaire:

- Qu. 1: Do you have children?  Yes □  No □.
  If you answered 'No' to Qu.1, go to Qu. 3.
- Qu. 2: Are any of your children aged 5 or under?  Yes □  No □.
- Qu. 3: ...

There are only three legitimate ways to answer these two questions, not four, as is the case with two independent questions. The type of my children has been truncated to the proposition as to whether I have children. A property which may be assessed for each child, *Being aged five or under*, now needs to be modified to a proposition relating to the single element of *I have children*. Here it is altered to some child having this age, but another question might have asked whether *all* are similarly youthful.

Now, if we take the special case where types have no genuine dependence, then forming dependent sum results in a product, $\sum_{x:A} B \simeq A \times B$. A product type such as *Day and hour* is such a dependent sum, since each day has twenty-four hours invariantly through the week and so $\sum_{d:day} Hour(d)$ collects pairs, such as (Monday, 15:00) and (Saturday, 09:00). In the case of propositions, this non-dependent dependent sum amounts to a conjunction, its element a proof of the first conjunct paired with a proof of the second. In a sense, then, conjunction is a degenerate form of 'and', and indeed it is typically inappropriate to use the word 'and' in such cases. On the other hand, with proper dependence, there is often a progressive flavour to the assertions.

- Pam took the key out of her bag and opened the door.

Does this imply that Pam used the key to unlock the door? You would certainly be surprised if, on being shown footage of the event, you saw that she took out the key, made no attempt to use it, but just pushed open the door, or used another key that happened to be in her hand. On the other hand, she might give off signs after withdrawing the key that she just recalled that she had in fact left the door unlocked. But then this would be mentioned in the report. The script strongly suggests the use of the key she took out of the bag. Indeed, it's reasonable to think the sentence above is elliptical for

- Pam took the key out of her bag and unlocked the door *with it*.

Now we see the dependence. We introduce a key, which is taken out of a bag and then used to enter, $k : Key, p : TakeOut(Pam, k, her\ bag), q : OpenWith(Pam, door, k),$

$$((k,(p,q)) : \sum_{x:Key} TakeOut(Pam, x, her\ bag) \& OpenWith(Pam, door, x).$$

We see then again the reliance on anaphoric reference, even if implicit, means that it makes no sense to ask of a dependent statement whether it is true, as it does in the case of separate conjoined propositions. If Jack did not in fact fall down the hill, then Jill cannot come tumbling after him. If there is no key for Pam to take out of her bag, then she cannot open her door with it.

Now, if 'and' demands linkages between parts, then '*but*' does so even more strongly. Here there is often a contrast between clauses or a denial of typical consequence:

- I take sugar in my tea, but sweeteners in my coffee.
- Phil tripped over, but he still won the race.

Suitable versions with '*and*' are:

- I take sugar in my tea and in my coffee.
- Phil tripped over and lost the race.

Similarly, *while, although, whereas, despite the fact that* are all forms of *and* with more stringent conditions:

- Jay likes beer, whereas Kay prefers wine.
- Despite the fact that Jane fell over, she still won the race.

Thus we have found that our '*and*', which appeared first to arise from the product of types, both the product of sets and the conjunction of propositions, has now taken on some extra refinement. Indeed, it seems to be used especially in cases where we have to bear in mind some aspect of what precedes the connective to tie it to some feature of what follows. Dependent sum allows us to represent precisely this.

Of course, not any thematic connection will work with '*and*'. As Rundle points out, we don't use it to join these sentences: 'You can't expect him to be here yet. The traffic is so heavy.' (1983, p. 389). This would be to run against the direction of any dependency. It is working backwards to give a reason for what ought to be expected. Viewing an explanation as the provision of premises from which the explanandum is derived, then we see there is a way to employ '*and*' here, namely, by saying 'The traffic is so heavy, and so you can't expect him to be here yet.'

Similarly, a challenging '*and*' is a request to hear what follows next.

- You haven't tidied your room yet. *And?* Well, I won't give you your pocket money then.

With these constructions we appear to have shifted away from the succession of pieces of information to the succession of some process of inference. One way to construe this is as an explicit proposition *A* and an implicit hypothetical $A \to B$ driving inference to *B*. When someone delivers you a loaded proposition, you can be sure there are consequences in the air.

But we might take there to be a dependency present again, for example,

- You haven't tidied your room and so I won't give you your pocket money as a punishment for not tidying your room.

Similarly, '*and*' appears explicitly in an implication in Winston Churchill's plea to the United States for arms in the Second World War:

- Give us the tools, and we will finish the job.

We might rephrase this:

- There is a job. If you give us the tools for this job, we will finish the job with these tools.

Recall that the reason I gave for starting this section with '*and*' was that it was thought to provide the logician the means to showcase their semantics for connectives. The reason for hesitancy about '*implies*' was the issue of the relevance of antecedent to consequent. Now that we have revealed that '*and*' is heavily mired in relevance considerations, yet emerges unscathed after the adoption of the dependent sum construction, perhaps we will find a satisfactory way to treat '*implies*' type-theoretically.

Earlier we saw that a proof of '*A* implies *B*' behaves like a function mapping proofs of *A* to proofs of *B*. Let's see what this says about the standard classical interpretation of the truth of $A \to B$ as *A* being false, *B* being true, or both. So if *A* is false, or in other words, if there are no proofs of *A*, then there is a function defined on this empty set, whatever the target. If this comes as news to the reader, it is the easiest function to define, since nothing needs to be specified.[9] On the other hand, if *B* is true and we have a proof, *p*, of it, then we can define a function on proofs of *A* as sending any of them to *p*. This is a constant function which ignores any information about *A*.

We see now where the Logic 101 student is surprised. 'If London is the capital of France, then *P*' turns out true, whatever *P*, because of this strange map from the empty set, and 'If *Q*, then Paris is the capital of France' turns out true, whatever *Q*, because we already know that Paris is the capital of France. The student senses that tricks are being played on them. What of genuine connections between the propositions?

Well, no stronger connection can be found than identity, so it is somewhat less surprising to be told of the truth of $A \to A$, whatever *A*. It may not be a particularly interesting result, but at least we have to rely on any proof we might have of *A* to send it to itself via the identity map. *A* may turn out to be false, but every eventuality is covered by the presence of that identity map. But 'if *P*, then *P*' is still a degenerate form of the use of this connective, even if

---

[9] For the category theorist, this is similar to the universal property of the empty set in the category of sets being the *initial object*, since there is a unique map from it to any object in the category.

it can play the kind of rhetoric function of 'If this is (indeed) what we must do, then this is what we must do'. Let us, then, consider some more substantial examples of the consequent relying on the antecedent:

- If you see something you like, then you should buy it.
- If we miss the last train, we will have to stay here for the night.
- If he says something to Jane and upsets her, I will be very cross with him.

Take the first of these. The type of things you like is a dependent sum, $\sum_{x:Thing} like(you, x)$, whose elements are pairs, each of which is a thing and evidence that you like it. Then we are being told that there's some function, $q$, which takes any instance of this type, say, $(a, r)$, with $a$ a thing, and $r$ a warrant that you like $a$, to a warrant, $q(a, r)$, for 'You should buy $a$'. As we will see, such a passage is provided by an element of a dependent product, denoted

$$q : \prod_{z:\sum_{x:Thing} like(you,x)} You \ should \ buy(p(z)),$$

where $p$ picks out $a$, the first component of $(a, r)$. It is possible that such a $q$ for '*you*' has been derived from a function defined on a collection of people, justifying their purchase of anything they like.

Just as dependent sum generalizes away from the cases of conjunction and product, so dependent product generalizes away from implication between non-dependent propositions and functions between sets, to maps between types and types dependent upon them. If the standard logical '*and*' was a degenerate form of dependent sum, the standard logical '*implies*' is a degenerate form of dependent product. The connective '*or*' does not generally arise in a similar way. We don't say 'Jack fell down or Jill came tumbling after' since the necessary dependency is not present, although we do use '*or*' in the form of an implication from the negation of a proposition. This is generally dependent as in 'This fine should be paid by midnight or it will be doubled.'

We need now to take a closer look at these dependent constructions, but before doing so let me draw a lesson from this section. In his *Stanford Encyclopedia of Philosophy* article, 'Moral Particularism', Jonathan Dancy (2017) uses the example of '*and*' to provide an instructive analogy of how moral reasoners cope with variation in the 'practical purport' of a concept such as *cruelty*:

> In knowing the semantic purport (= the meaning) of 'and', one is in command of a range of contributions that 'and' can make to sentences in which it occurs. There need be no 'core meaning' to 'and'; it would be wrong to suggest that 'and' basically signifies conjunction. If you only know about conjunction, you are not a competent user of 'and' in English, for there are lots of uses that have little or nothing to do with conjunction. For example: two and two make four; 'And what do you think you are doing? (said on discovering a child playing downstairs in the middle of the night); John and Mary lifted the boulder; the smoke rose higher and higher. Those competent with 'and' are not unsettled by instances such as these, but nor are they trying to understand them in terms of similarity to a supposed conjunctive paradigm or core case (Dancy 2017).

Dancy points to a 'manageable complexity' in this case, and claims this to be so in moral cases, too. From what we have seen we can take the manageability of the variation in 'and' to be explained by the limited range of ways it can be taken up by our type theory. For instance, 'The smoke rose higher and higher' would seem to employ 'and' with its sense of dependency so that it might be written less briefly as: *The smoke rose higher than it had been, and then higher again than that.*

## 2.4  Dependent Types

Dependent types are a very important part of Martin-Löf type theory. As we saw, these are denoted $x : A \vdash B(x) : Type$, where the type $B(x)$ *depends* on an element of $A$. In mathematics we may have a collection of types indexed by the natural numbers, such as the type of $n \times n$ matrices over the real numbers, $n : \mathbb{N} \vdash Mat(n, \mathbb{R}) : Type$. But let us work with an example from ordinary language, as in *Players(t)* for $t : Team$. So we express this as

$$t : Team \vdash Player(t) : Type.$$

These dependent types are sets, in a sense to be defined later, but we can have examples where they are propositions, such as

$$t : Team \vdash Plays\ in\ UK(t).$$

Quantification then takes place in these dependent-type situations, where we find that domains of variation are the indexing types. This relies on the type formation rules for *dependent sum* and *dependent product*. For the dependent type $B(x)$ depending on $x : A$,

- The dependent sum (sometimes known as dependent *pair*), $\sum_{x:A} B(x)$, is the collection of pairs $(a, b)$ with $a : A$ and $b : B(a)$. When $A$ is a set and $B(x)$ is a constant set, $B$, this construction amounts to the product of the sets. Likewise, if $A$ is a proposition and $B(x)$ is an independent proposition, $B$, dependent sum is the conjunction of $A$ and $B$. In general we can think of this dependent sum as sitting 'fibred' above the base type $A$, as one might imagine the collection of league players lined up in fibres above their team name.

- The dependent product (sometimes known as dependent *function*), $\prod_{x:A} B(x)$, is the collection of functions, $f$, such that $f(a) : B(a)$. Such a function is picking out one element from each fibre. When $A$ is a set and $B(x)$ is a constant set, $B$, this construction amounts to $B^A$, the set of functions from $A$ to $B$. Likewise, if $A$ is a proposition and $B(x)$ is an independent proposition, $B$, dependent product is the implication $A \rightarrow B$. In terms of the picture of players in fibres over their teams, an element of the dependent product is a choice of a player from each team, such as *Captain(t)*.

**Table 2.1** Dependent sum and dependent product.

| Dependent sum | Dependent product |
|---|---|
| $\sum_{x:A} B(x)$ is the collection of pairs $(a, b)$ with $a : A$ and $b : B(a)$. | $\prod_{x:A} B(x)$, is the collection of functions, $f$, such that $f(a) : B(a)$. |
| When $A$ is a set and $B(x)$ is a constant set, $B$: The product of the sets. | When $A$ is a set and $B(x)$ is a constant set, $B$: The set of functions from $A$ to $B$. |
| When $A$ is a proposition and $B(x)$ is an independent proposition, $B$: The conjunction of $A$ and $B$. | When $A$ is a proposition and $B(x)$ is an independent proposition, $B$: The implication $A \to B$. |

Here, then, we can see how close this foundational language is to mainstream mathematics and physics. Dependent sum and dependent product correspond respectively to the total space and to the space of sections of fibre bundles, which appear in gauge field theory often in the guise of principal bundles. A fibre bundle is a form of product, but a potentially *twisted* one, where as we pass around the base space a fibre may be identified under a nontrivial equivalence. An easy example is the Moebius strip, where an interval is given a 180-degree twist as it is transported around a circle. Of course, for physics one needs some geometric structure on the base space and total space. We shall see more about this in Chap. 5.

Quantification is associated to the second example in each case where the dependent type is a proposition. The dependent sum being inhabited amounts to the existence of a team that plays in the UK, and the dependent product being inhabited amounts to all teams playing in the UK. We can now see that these constructions allow us to formulate the quantifiers from first-order logic, at least as defined over types. So consider the case where $A$ is a set, and $B(a)$ is a proposition for each $a$ in $A$. Perhaps $A$ is the set of animals and $B(a)$ states that a particular animal, $a$, is bilateral. Then an element of the dependent sum is an element $a$ of $A$ and a proof of $B(a)$, so something witnessing that there is a bilateral animal. Meanwhile, an element of the dependent product is a mapping from each $a : A$ to a proof of $B(a)$. There will only be such a mapping if $B(a)$ is true for each $a$. If this were the case, we would have a proof of the universal statement 'for all $x$ in $A$, $B(x)$', or, in our example, 'All animals are bilateral'.

Returning to the dependent sum, this is almost expressing the usual existential quantifier 'there exists $x$ in $A$ such that $B(x)$', except that it's gathering all such $a$ for which $B(a)$ holds, or, in our case, gathering all bilateral animals. As we have seen before in the capped addition of the Boolean truth values in a disjunction, to treat this dependent sum as a proposition, there needs to be a 'truncation' from set to proposition, so that we ask merely whether this set is inhabited, or, in our case 'Does there exist a bilateral animal?' This extra step should be expected, as existential quantification resembles forming a long disjunction. That we don't need to adapt for universal quantification tallies with the straightforward form of the product of Boolean values.

What emerges from this line of thought is that the lower levels of homotopy type theory have contained within them: propositional logic, (typed) predicate logic and a structural set theory. Coming from a tradition of *constructive* type theory, one needs to add classical axioms if these are required, such as various forms of excluded middle or axioms of choice.

The difference with an untyped setting is very apparent when we look to express something with multiple quantifiers, such as 'Everyone sometimes finds themselves somewhere they don't want to be'. In type theory there will be dependency here separately on types of people, times and places, and not variation over some universal domain, requiring conditions that specify that some entities in the domain be people, times or places.

We saw a simplified version of Göran Sundholm's resolution of the puzzle of the Donkey sentence in Chap. 1. A treatment of the full version should be illuminating now, since it combines the constructions we have just seen (Sundholm 1986):

- If a farmer owns a donkey, then he beats it.

Recall that the problem here is that we expect there to be a compositional account of the meaning of this sentence, in particular, one where the final 'it' appears in the representation. It appears that an existential quantifier would be involved in a representation in first-order logic because of the indefinite article, and yet a simple-minded attempt to use one is ill-formed, the final $y$ being unbound:

$$\forall x(Farmer(x)\&\exists y(Donkey(y)\&Owns(x,y)) \rightarrow Beats(x,y)).$$

It is standard, then, in first-order logic to rephrase the sentence as something like: 'All farmers beat any donkey that they own', and then to render it formally as

$$\forall x(Farmer(x) \rightarrow \forall y(Donkey(y)\&Owns(x,y) \rightarrow Beats(x,y))).$$

But now we have radically transformed the original sentence, and the 'it' does not seem apparent.

Sundholm's solution uses the resources of dependent type theory:

- $\vdash Farmer : Type,$
- $\vdash Donkey : Type,$
- $x : Farmer, y : Donkey \vdash own(x,y), beat(x,y) : Prop,$
- $x : Farmer \vdash Donkey\ owned\ by\ x : Type,$
- $\vdash (\sum_{x:Farmer} Donkey\ owned\ by\ x) : Type.$

Then the statement is true if we have a function, $b$, such that

- $z : (\sum_{x:Farmer} Donkey\ owned\ by\ x) \vdash b(z) : Beats(p(z), p(q(z))).$

Elements of the dependent sum of donkeys owned by farmers are pairs formed of a farmer and then a pair formed of a donkey and a warrant that the donkey is owned by that farmer.

From such an element, $z$, we project to the first component of the pair, $p(z)$, to extract the farmer and then project to the first component of the second component, $p(q(z))$, for the donkey, so as to be able to express the beating of one by the other. It is this last term that is being referred to in natural language as 'it'.

The $b$ above provides a proof of the relevant beating for any such $z$ and, as such, is then an element of the following dependent product:

$$b : \prod(z : (\sum(x : Farmer)\sum(y : Donkey)Owns(x,y)))Beats(p(z),p(q(z))).$$

Now, sometimes the referents of these anaphoric pronouns are not so obvious. Consider the song 'If You're Happy and you Know it, Clap your Hands'. What does the 'it' refer to in the antecedent?

- You're happy and you know it.

Following Vendler, who claimed that we *believe* propositions but *know* facts, we might say this is

- You're happy and you know the fact that you're happy.

So how is '$X$ knows $P$' as a type formed? Well, for $P$ to be known it had better be true, so we can make a precondition that the type which is $P$ is inhabited:

$$X : Person, P : Prop, x : P \vdash know(X,P,x) : Type.$$

We might take this in turn to be a proposition if we consider the singular fact of $X$ knowing a true $P$, without distinguishing variety of warrants for ascribing this knowledge to $X$. Typically, in natural language we merely write '$X$ knows $P$' without indicating $P$'s element. But with the implicit dependence on $x$, we can say that the formation of this proposition *presupposes* the truth of $P$, in a sense that will be explained in §2.5. Our epistemology will dictate here a complex set of requirements for judging a knowledge assertion, likely dependent on the kind of proposition $P$ at stake.

According to this way of framing matters, knowing that one knows becomes quite involved, and the KK principle (that knowing implies knowing that one knows) won't hold. To form '$X$ knows that $X$ knows $P$' we need not only that $P$ be true, but also that it's true that $X$ knows $P$:

$$P : Prop, x : P, y : Know(X,P,x) \vdash Know(X,Know(X,P,x),y) : Prop.$$

To establish this latter proposition as true, we will need further to demonstrate that this latter type is inhabited, and this will now depend on the kind of proposition at stake and the kind of warrant for its being known.

Once we have the judgements:

- ⊢ *you* : *Person*,
- ⊢ *you're happy* : *Prop*,
- ⊢ *h* : *you're happy*,
- then, ⊢ *know(you, you're happy, h)* : *Prop*
- and we may have ⊢ *k* : *know(you, you're happy, h)*,
- and so ⊢ $(h, k)$ : $\sum_{x:You're\ happy} know(you, you're\ happy, x)$.

The 'and' of '*You're happy and you know it*' has again been represented by a dependent sum. The 'it' is the pair ( *you're happy, h*). In accordance with Vendler, we might contrast

$$X : Person, P : Proposition, x : P \vdash know(X, P, x) : Type$$

with

$$X : Person, P : Proposition \vdash believe(X, P) : Type.$$

So we don't sing 'If you're happy and you believe it, clap your hands.' Of course, 'I believe it (her claim)' is fine, and the claim may in fact be false.

A very similar and much more extensive treatment of this believe/know distinction in terms of dependent type theory is given in Tanaka et al. (2017). The authors note in an earlier version of the paper (Tanaka et al. 2015) that the distinction is lexically marked in Japanese. For example, in the case of knowledge

- John-wa Mary-ga kita *koto*-o sitteiru.
- John-TOP Mary-NOM came COMP-ACC know.
- 'John knows (the fact) that Mary came.'

In contrast, for belief

- John-wa Mary-ga kita *to* sinziteiru.
- John-TOP Mary-NOM came COMP believe.
- 'John believes that Mary came.'

They continue

In general, *koto*-clauses trigger factive presupposition, while *to*-clauses do not. This contrast can be captured by assuming that a factive verb like *sitteiru* takes as its object a proof (evidence) of the proposition expressed by a *koto*-clause, while a non-factive verb selects a proposition denoted by a *to*-clause. (Tanaka et al. 2015, p. 6)

Before ending this section, in the spirit of *computational trinitarianism*, I shall first give the type-theoretic rules for dependent sum and product, and then give a category-theoretic account of their universal properties in terms of adjoints. First, then, dependent sum:

$$\frac{\vdash X : Type \quad x : X \vdash A(x) : Type}{\vdash \left(\sum_{x:X} A(x)\right) : Type}$$

$$\frac{x : X \vdash a(x) : A(x)}{\vdash (x, a) : \sum_{x':X} A(x')}$$

$$\frac{t : \sum_{x:X} A(x)}{p_1(t) : X \quad p_2(t) : A(p_1(t))}$$

$$p_1(x, a) = x \quad p_2(x, a) = a.$$

Then the rules for dependent product:

$$\frac{\vdash X : Type \quad x : X \vdash A(x) : Type}{\vdash \left(\prod_{x:X} A(x)\right) : Type}$$

$$\frac{x : X \vdash a(x) : A(x)}{\vdash (x \mapsto a(x)) : \prod_{x':X} A(x')}$$

$$\frac{f : \prod_{x:X} A(x) \quad x : X}{x : X \vdash f(x) : A(x)}$$

$$(y \mapsto a(y))(x) = a(x).$$

Recall from above that a logical inferentialist such as Boghossian understands the meaning of the conjunction '*and*' to be given by its introduction and elimination rules. With our more adequate treatment of '*and*' as dependent sum for propositions, we can maintain a similar stance, but now with these new versions of the rules.

Let's now look at the category-theoretic understanding, presented as usual in terms of *adjoints*. It should not be surprising to find that dependent sum and product can be formulated in terms of adjunctions in view of a similar treatment for conjunction and implication shown in §2.2. As I mentioned there, adjunctions are a vital part of category theory, and provide a way of dealing with something as close to an inverse as possible. When such an inverse does not exist, left and right adjoints are approximations from two sides.

In our type theory, given a plain type, $A$, we can turn any type, $C$, into one trivially dependent on $A$ by formulating $x : A \vdash (A \times C)(x) :\equiv C : Type$. If $A$ and $C$ are sets, think of lining up a copy of $C$ over every element of $A$, the product of the two sets projecting down to the first of them, $A \times C \to A$. Thus we have made $C$ depend on $A$, but in a degenerate sense where there is no real dependency. Now, we can think of approximating an inverse to this process, which would need to send $A$-dependent types to plain types. Such approximations, or *adjoints*, do indeed exist. Left adjoint to this mapping is dependent sum, and right adjoint is dependent product. The fact that these are adjoints may be rendered as follows for $B$, a type depending on $A$:

- $Hom_{\mathcal{C}}(\sum_{x:A} B, C) \cong Hom_{\mathcal{C}/A}(B, A \times C)$
- $Hom_{\mathcal{C}}(C, \prod_{x:A} B) \cong Hom_{\mathcal{C}/A}(A \times C, B)$.

$\mathcal{C}/A$ indicates the slice of $A$. Objects in this slice are objects of $\mathcal{C}$ equipped with a mapping to $A$. A morphism between two such objects is a commutive triangle. These correspond to the $A$-dependent types. Think of a couple of types dependent on $t : Team$, the type of teams; say, $Player(t)$ and $Supporter(t)$. Then an example of a map from $Supporter$ to $Player$ in the slice over $Team$ is *favourite player*, assuming that each supporter's favourite player belongs to the team that they support.

This category-theoretic formulation of dependent sum and product will reappear in Chap. 4 when a type of possible worlds comes to play the role of $A$. Let me end here by noting that, in light of the adjunctions above, we can see further developments of the logic–arithmetic analogy from earlier in this chapter. Take $A$, $B$ and $C$ as finite sets, with $B$ fibred over $A$. This just amounts to a map from $B$ to $A$, the elements of $B$ fibred above their images in $A$. Let $a$ be the cardinality of $A$, $c$ be the cardinality of $C$ and $b_i$ be the cardinality of the subset of $B$ which is the fibre over $i$ in $A$. Then, taking the cardinalities of the two sides of each adjunction above yields further recognizable arithmetic truths:

- $c^{\sum_i b_i} = \prod_i c^{b_i}$
- $(\prod_i b_i)^c = \prod_i (b_i)^c$.

So a pupil being taught, say, that $3^4 \times 3^5 = 3^9$ or that $3^5 \times 7^5 = 21^5$ is being exposed to the shadows of instances of important adjunctions, which, in turn, as with all of the discussion above of dependent sums and products, works for types up and down the hierarchy of $n$-types. These $n$-types in the guise of higher groupoids we turn to in Chap. 3. Let us now consider the nature of the dependence relation between types.

## 2.5 Context and Dependency Structure

Consider what might be the beginning of a story, or a play:

> A man walks into a bar. He's whistling a tune. A woman sits at a table in the bar. She's nursing a drink. On hearing the tune, she jumps up, knocking over the drink. She hurls the glass at him. 'Is that any way to greet your husband?' he says.

For Ranta (1994), this kind of narrative should be treated as the extension of one long *context*, with its dependency structure, which begins as follows:

> $x_1 : Man, x_2 : Bar, x_3 : WalksInto(x_1, x_2), x_4 : Tune, x_5 : Whistle(x_1, x_4), x_6 :$
> $Woman, x_7 : Table, x_8 : Locate(x_7, x_2), x_9 : SitsAt(x_6, x_7), x_{10} : Drink, x_{11} :$
> $Nurse(x_6, x_{10}), x_{12} : Hear(x_6, x_5), \ldots$

Note how one could easily populate this sketch with plenty of 'and's, especially where there is dependency, suggesting that there are dependent sums about.

In general, a context in type theory takes the form

$$\Gamma = x_0 : A_0, x_1 : A_1(x_0), x_2 : A_2(x_0, x_1), \ldots, x_n : A_n(x_0, \ldots, x_{n-1}),$$

where the $A_i$ are types which may be legitimately formed. As we add an item to a context, there may be dependence on any of the previous variables. A context need not take full advantage of this array of dependencies. For instance, in the case above, *Whistle* only depends upon $x_1$ and $x_4$ and not upon $x_2$ or $x_3$. But it certainly cannot depend on a variable ahead of it.

The idea of a context can help us make sense of R G Collingwood's theory of *presuppositions*, described in his *An Essay on Metaphysics* (1940). There he argues against the idea that there can be freestanding propositions. Rather, any statement is made as an answer to a question. That question, in turn, relies on some further presuppositions. These can be traced back to further questions, until one reaches the bedrock of *absolute* presuppositions. These presuppositions are not truth-apt, that is, are not to be assessed for their truth value.

> Absolute presuppositions are not verifiable. This does not mean that we should like to verify them but are not able to; it means that the idea of verification is an idea which does not apply to them, because, as I have already said, to speak of verifying a presupposition involves supposing that it is a relative presupposition. If anybody says 'Then they can't be of much use in science', the answer is that their use in science is their logical efficacy, and that the logical efficacy of a supposition does not depend on its being verifiable, because it does not depend on its being true: it depends only on its being supposed. (1940, p. 32)

It is possible, however, for the *constellation* of absolute presuppositions of a science to change. Their *logical efficacy* may be found wanting, relative to a different constellation.

Collingwood (1940, p. 38) illustrates the role of presuppositions in an everyday situation by considering the 'complex question' *Have you left off beating your wife?* Notoriously, to answer 'yes' or 'no' to this question exposes you to censure, and yet what are you to do if either you have no wife, or you have one but have never beaten her? Collingwood's response is to say that questions only arise in certain circumstances, and to find out these circumstances it is necessary to work out a series of questions to which a positive answer allows the next question. He arrives at the following series:

1. Have you a wife?
2. Were you ever in the habit of beating her?
3. Do you intend to manage in the future without doing so?
4. Have you begun carrying out that intention?

Similar to the opening narrative of this section, we can tell a simple story here.

A man and a woman met and got married. At some stage in their marriage, he began to beat her. In time he came to see this had to stop, and so he formed the intention to desist. He acted on this decision and to this day has abstained.

As I mentioned, Collingwood claimed that his method, which he called the *logic of question and answer*,[10] would allow us to reveal the *absolute presuppositions* in operation in some walk of life. The work of the metaphysician is to perform just this kind of revelation for bodies of organized thought, '*sciences*' as he calls them, at different epochs. This is to allow us to see how our deepest foundational concepts, such as *cause*, have been transformed through time. Even in the everyday case we just considered, the analysis might clearly probe deeper. Imagine the child pestering its parents with a series of questions.

- *What is a wife?* A woman to whom one is married. *What is marriage?* When two people agree to a legally binding union. *What is it for two people to agree?* ...

Type theoretically, all of the uncovered presuppositions need to appear in the context right up to the formation of the type *Person*.

Type theory, with its resources of context and type formation, may allow us better to keep track of our background theoretical assumptions. For instance, as I encounter different jurisdictions, I must keep a tally of the regulations concerning marriage. The rules of marriage are ever-evolving, often in contested ways, both in terms of the conditions that need to be in place for it to be recognized that two people be married, and in terms of what follows morally, legally, financially, and so on, from the marriage. We have seen substantial changes over the past few decades, and should only expect more to follow, such as perhaps to the presumption that only two people may be married, as put in question by the polyamory community.

Another philosopher arguing along similar lines to Collingwood in the context of the natural sciences is Michael Friedman. Friedman (2001) contrasts his own views with those of Quine, who famously proposed that we operate with a connected web of beliefs (Quine and Ullian 1970). For the latter, when observations are made which run against expectations derived from these beliefs, we typically modify peripheral ones, allowing us to make minimal modification to the web. However, we may become inclined to make more radical changes to *entrenched* beliefs located at the heart of the web, perhaps even to the laws of arithmetic or the logic we employ. Friedman insists, by contrast, that there is a hierarchical structure to our (scientific) beliefs. This entails that, without the availability of fundamental modes of expression, including the resources of mathematical languages, and constitutive principles which deploy these resources, statements concerning empirical observations and predictions cannot even be meaningfully formed. In his (2002), while discussing the formulation of Newtonian physics, he says

It follows that without the Newtonian laws of motion Newton's theory of gravitation would not even make empirical sense, let alone give a correct account of the empirical phenomena: in the absence of these laws we would simply have no idea what the

---

[10] See also Collingwood (1939, Chap. 5).

relevant frame of reference might be in relation to which the universal accelerations due to gravity are defined. Once again, Newton's mechanics and gravitational physics are not happily viewed as symmetrically functioning elements of a larger conjunction: the former is rather a necessary part of the language or conceptual framework within which alone the latter makes empirical sense. (Friedman 2002, pp. 178–9)

To see how similar this case is to that of the beaten spouse, we might imitate Friedman thus:

It follows that without the concepts of personhood, of marriage, of action, of intention, etc., the idea of someone leaving off beating their wife would not make any sense, let alone give a correct account of the empirical phenomena: in the absence of these concepts, we would simply have no idea what the relevant moral–legal–ontological frame of reference might be in relation to which the claimed cessation of violence has taken place. Once again, facts about personhood, intention and action are not happily viewed as symmetrically functioning elements of a larger conjunction: the former is rather a necessary part of the language or conceptual framework within which alone the latter makes empirical sense.

What neither Collingwood nor Friedman went on to do, however, is to give an appropriate formal treatment of this dependency. Let us see how we fare with dependent type theory.

With the idea of a context available, type theory should provide us with a way to distinguish between *consequences* and *presuppositions*. Whereas 'James has stopped smoking' *presupposes* that at some point he was in the habit of smoking, from the proposition '$n$ is the sum of two odd numbers' we may *conclude* that $n$ is even. Presuppositions are what need to be in place for a type to be formed or a term introduced. Consequences result from application of rules to existing judgements. For example,

- Presupposition: from $A \times B : Type$, we must have that $A, B : Type$.
- Consequence: from $c : A \times B$, we can conclude that $p(c) : A$ and $q(c) : B$.

Despite this neat distinction, we are inclined to see an inference occurring in the case of presupposition just as much as in the case of consequence. It is easy to think that we infer from 'James has stopped smoking' that he was in the habit of smoking. What seems to be occurring is that we are presented with a proposition, $P$, such as 'James has stopped smoking'. Then, to interpret it, it is our job to provide the minimal context, $\Gamma$, which can support the formation of $P$:

$$\Gamma \vdash P : Prop.$$

Then say $\Gamma = x_1 : A_1, x_2 : A_2, ..., x_n : A_n$, from which we extract one clause of the context

$$x_1 : A_1, x_2 : A_2, ..., x_n : A_n \vdash x_j : A_j, for \ 1 \leq j \leq n.$$

We have trivially

$$x_1 : A_1, x_2 : A_2, ..., x_n : A_n, y : P \vdash x_j : A_j,$$

at which point we forget the context and think we have an entailment

$$y : P \vdash x_j : A_j.$$

This is to be contrasted with the case when we have the means in a context to derive a new proposition, such as with

$$x : A, f : [A, B] \vdash f(x) : B.$$

Ranta in 'Constructive Type Theory' (2015, p. 358) proposes along similar lines that we distinguish these operations as follows:

- $B$ presupposes $A$ means $x : A \vdash B(x) : Type$.
- $A$ semantically entails $B$ means $x : A \vdash b(x) : B$.

The centuries-old disagreement as to the status of *Cogito ergo sum* would appear to be relevant here. Where Carnap in 'The Elimination of Metaphysics' (1932) takes Descartes to task for a faulty piece of deductive reasoning, one might rather say that a presupposition is being revealed – *I think* presupposes *I am*. Heidegger observes

> Descartes himself emphasizes that no inference is present. The *sum* is not a consequence of thinking, but vice versa; it is the ground of thinking, the fundamentum. (Heidegger 1967, pp. 278–9)

Perhaps the 'I', rather than referring to an element of a type, plays more of a performative or expressive role in our assertions,[11] as represented in the judgement symbol, ($\vdash$), but a third-person proposition such as *Jill thinks* certainly presupposes that there is a *Jill* who is one of a kind that can think.

Certainly worth exploring here is whether forms of inferentialism could be well-represented type theoretically, as above. In the case of Brandom's version of inferentialism with his idea of *commitments* and *entitlements*, see, for example, *Making it Explicit* (Brandom 1994), we see a connection not only with the pairing of introduction and elimination rules, but also with the distinction between constructing implicit context and calculating consequents. There, we are told, to understand the meaning of a proposition (something approximating what we are calling a judgement in the sense of a thing to be judged, rather than the act of judging) we should understand what conditions entitle us to

---

[11] See 'The "emptiest of all representations," the "'I think' that can accompany all representations" expresses the formal dimension of responsibility *for* judgments' (Brandom 2000, p. 160).

assert it, as well as what follows from asserting it. Perhaps in most forms of not fully spelled out inference, what we are 'making explicit' is the context.

With such a rich notion of context, we may also be able to provide support for a philosopher of language such as Jason Stanley in *Language in Context* (2007), who believes that contextualists have overstated the case, and that reliance on the context of utterance is limited to indexical aspects. Clearly, when it comes to indexicals, we do rely on *situational* context. Where I point is evidently vital to establishing the truth of the claim 'This is a kangaroo'. Stanley looks to limit the role of this situational context to the indexical aspects of an assertion. The remainder is borne by what we might call the *conceptual* context of the utterance. Where he sees his task as 'attributing hidden complexity to sentences of natural language, complexity which is ultimately revealed by empirical enquiry' (2007, p. 31), we may describe this as revealing all aspects of the type-theoretic context.

A further consideration will be to analyse ill-formed constructions. We will look at the case of ill-formed definite description in the following chapter, but let us consider briefly here the attempt to generate the proposition which supposedly leads to the liar paradox:

- *This proposition is false.*

Now, a reasonable rule for forming terms of the kind 'this $A$' for some established type $A$ is that we have just presented a term of that type, say, $a : A$. Then we may define

$$\vdash this\, A :\equiv a : A.$$

But in the case of the liar sentence, we haven't yet produced a proposition, an element of the type *Prop*, in order to be able to form 'this proposition'. In other words, the liar statement presupposes that a proposition has just been presented, but we don't yet have one. Without it, we cannot introduce the term 'this proposition'. Once again, we find an answer which follows the Wittgensteinian line, but we can present it formally.

Let us note one final feature in Collingwood's wife-beating example—the choice of the perfect tense. This tense requires that at the time of asking the question, an affirmative answer indicates that the intention to stop is still being adhered to. The English perfect always has this connotation of present relevance. Mention of complex temporally structured events brings us back to an outstanding topic which we can address now we have the tools provided by dependent types.

## 2.6 Events as Basic Types

Martin-Löf type theory was designed to be a language for expressing mathematical reasoning. Very few basic types need to be postulated, since most can be built up from the type formation rules. In a type theory which allows for *inductive types* and also *higher* inductive types, described in Chap. 3, we have a type of natural numbers, and can then build up rationals and real numbers in something like the usual way. We can also form quotients of existing types and types with more complicated identity structures, such as $n$-spheres for all $n$ and algebraic entities.

However, if we hope to develop type theory to capture ordinary language and everyday inference, we will have to assert that there are some types composed of elements from our world.[12] We considered in the first section of this chapter the prospect of employing a type of entities or individuals, and then extracting out of it subtypes of entities of different kinds. This follows the current standard practice of representing quantifiers over restricted domains by including conditions: for instance, $\forall x(A(x) \rightarrow B(x))$ for 'All $A$ are $B$'. I provided some reasons for refusing to adopt this approach and instead taking there to be a number of basic types. Now that we have available the tools to represent dependency, we can say more.

Among the active research groups working on the representation of natural language by dependent type theory, we find both the position that each common noun denotes a type (Luo 2012) and the rival position that these nouns should be formed by predication on some master type *Entity* (Tanaka et al. 2015). One reason we might look to the latter is that otherwise a simple proposition such as '*John is a man*' presents problems. Since this is a proposition, and can be negated and appear in conditionals, it needs to be represented as a type. This would seem to rule out the syntactical form $John : Man$, since in type theory this is not a proposition. We might decide instead, then, to use $Man(x)$ as a predicate defined for some master type of entities, so that 'John is a man' is rendered as $(John, r) : \sum_{x:Entity} Man(x)$, where $r : Man(John)$. Then 'John is not a man' is the proposition $\neg Man(John)$.

Now, Luo and colleagues have formulated solutions to this challenge within the 'each common noun denotes a type' paradigm.[13] But even were we to find these solutions unconvincing and consider Tanaka to be closer to the mark, there is good reason to resist ascent to too high-level a master type. Let us see why, by considering how our sentence— *John is a man*—could come to be asserted, denied or included within a conditional.

First off, the interlocuter of someone making such a claim about 'John' must be expected to know something about the referent of 'John' prior to its assertion. For it to be meaningful to assert or deny the humanness of John, it has to be thinkable already that John might not be a man, so he cannot have been initially presented as such. By way of comparison, it is difficult to make sense of '*If red is not a colour, then I will wear a blue shirt*', since it seems we could not have been introduced to *red* other than as a colour. On the other hand, for a subtype of *Colour* this is fine, as in '*If red is not a colour that you like, then I will wear a blue shirt*'. It seems, then, that if '*John is a man*' is to make sense as an assertion, we would need some context as to how the term *John* has been introduced. Perhaps the speaker has explained how John was working in the fields all of yesterday, that John pulled heavy loads and that John was given just a bowl of oats in the evening. We might reply '*If John is a man, then you're not treating him well.*' We have accepted *John* as a something, and are wondering about predication as a man. However, even here, instead of some master type of individuals or entities, there is good reason to opt for separate basic types.

[12] These may be termed 'kinds'. 'Types are on the side of mind, kinds are on the side of the world' (Harré, Aronson and Way 1994, p. 27).
[13] See Chatzikyriakidis and Luo (2017), Xue et al. (2018).

Note that although we do not take common nouns to be types, it is possible to refine type **entity** by introducing more fine-grained types such as ones representing animate/inanimate objects, physical/abstract objects, events/states, and so on. (Tanaka et al. 2015, p. 3)

The characterizations given of John strongly suggest at the very least that he is an animate being. John cannot be referring to, say, a meteorological event.

But don't we sometimes hear a term and then labour under some misapprehension as to its type? I recall once embarrassing myself by speaking of the *hermeneutic circle* as though it were a group of intellectuals such as the *Vienna Circle*. Shouldn't we say that I knew that the term had singled out an individual, typed to a very minimal extent such that greater specification could class it under 'processes to understand texts' or 'groups of intellectuals'? And, as far as I knew, perhaps it referred to a geometric figure. All I knew confidently was that it purported to pick out some individual entity.

Indeed, I might mutter to myself 'If the hermeneutic circle is not a group of intellectuals, then I have made a complete fool of myself', and I could replace *group of intellectuals* with any number of types, and this sentence still makes sense. One way to think about this situation, however, is that this comment is taking place at the *meta*-level. I had assumed as a part of the common context of conversation the judgement that *hermeneutic circle* was an element of the type *Intellectual group*, but I was wrong to do so. My muttering in full is 'If I was wrong to assume the judgement that the hermeneutic circle is a group of intellectuals, then I have made a fool of myself.'

With this preamble over, let's see if we can say more about the type *Event*, a type with an intricate internal structure and with a reasonable claim to be basic. Those, such as Peter Hacker, dubious of the possibility of formalization as a tool to reveal meaning, have taken events as an important case in support of their views:

> The ideal of displaying the meaning, in particular the entailments, of sentences about events as wholly or even largely a function of structure as displayed in the canonical notation of the predicate calculus is chimerical. (Hacker 1982, p. 485)

Hacker is correct to draw our attention to the kinds of inference we need to be able to represent with our formalism, and in his claim that we need to fare better than is possible with the predicate calculus. If we can show that it is possible to draw inferences from complex event statements, and if the inference is formalizable in type theory, it must be because there is an implicit type structure operating behind the scenes. One is either inferring presuppositions, deducing consequences or both. So from

- Meg has wiped all the crumbs off the table,

we can reasonably conclude amongst other things that

- Meg moved her hand in a sweeping motion.
- Meg was in contact with the table.

- There were crumbs on the table.
- There are no longer crumbs on the table.

Now, the reasons Donald Davidson gives for postulating that events make up their own ontological category include that, on the one hand, as with objects, descriptive details can be added endlessly, and on the other, they have their own distinct form of identity criterion which differs from the identity criterion for objects. Let us take these in order.

When I consider a particular book in my office, I see that it has a blue cover, was written by Jane Austen, is in contact with a work by Charles Dickens. I know that it was given to me by my wife, we read it together last year, and that it contains a story about the difficulties of finding a suitable husband in the years around 1800. Further investigation reveals it has 346 pages, including an introduction of 24 pages, and 61 chapters. So now you have several details about the book, but I have only just begun. I haven't told you yet about its production method or its publication details. I haven't mentioned the font used in the text or the design of the frontispiece, and so on.

Now prima facie this description does not cause a problem for a predicate calculus representation. Taking the constant $a$ to designate the item, we have

$$Book(a)\&BlueCover(a)\&WrittenAusten(a)\&...,$$

which allows us to infer any part of the conjunction, such as

$$Book(a)\&WrittenAusten(a).$$

Davidson noted that a similar device could be used with events. Let us modify an example devised by Peter Hacker:

- *P*: The stone hit the car at the crossroads at midnight with a bang.

This might be rendered by a relation as

$$hit(the\ stone, the\ car, at\ the\ crossroads, at\ midnight, with\ a\ bang).$$

However, the problem Davidson raised with such an elaborate description of an event is that it seems as though we need related predicates of different *arities*. Simple consequences of the proposition are: the stone hit the car; the stone hit the car at midnight; the stone hit the car at the crossroads at midnight; the car was hit at midnight. If we imagine that there is a five-place predicate, $hit(u, v, x, y, z)$, expressing

- $u$ hit $v$ at location $x$ at time $y$ in manner $z$,

then either we also have to have predicates of other arities such as $hit$ (*that stone, the car*), $hit$ (*that stone, the car, midnight, with a bang*), and so on, or else we allow dummy placeholders so that 'the stone hit the car' is $hit$ (*the stone, the car*, ∗, ∗, ∗).

The former solution is enormously profligate, each predicate occurring in a great number of forms. As for the latter solution, it seems implausible that we conceive of the *hit* predicate in this way, with many of its instances having unfilled slots. In any case, as Hacker suggests, there is nothing to prevent us extending the description by adding extra details to the event: for instance, that the stone hit obliquely.

Davidson's solution was to imitate our treatment of the book above and so allow quantification over events

$$P :\equiv \exists e(hitting(e) \ \& \ agent(e) = the \ stone \ \& \ object(e) = the \ car$$
$$\& \ place(e) = the \ crossroads \ \& \ time(e) = at \ midnight$$
$$\& \ manner(e) = with \ a \ bang).$$

This solution allows for the inference to the reduced descriptions by removing the relevant conjuncts. So we may infer from the proposition, for instance, that 'the stone hit the car at the crossroads'.

This solution might suggest that we take events along with objects as falling under the larger class of *individuals*, so, for example,

- $\exists x(object(x) \ \& \ book(x) \ \&...$
- $\exists x(event(x) \ \& \ hitting(x) \ \&...$

But objects and events have such different individuation criteria that Davidson preferred the solution which distinguishes them as of fundamentally different kinds. His identity criterion for events is:

- Events are the same if they have the same causes and are caused by the same things.

Hacker raises several objections, however, to Davidson's account. First, he observes,

If the stone hit the car then surely, as a matter of logic, the stone and the car must exist. (1982, p. 484)

Of course, he's right that these are not derivable from $P$; however, they are *presupposed* by it. Rather like the liar sentence case treated above with its use of 'this', and as we shall see in greater detail in Chap. 3, we cannot form sentence $P$ with its use of the definite article, 'the', without having introduced an element of the type *Stone* and an element of the type *Car*. We will already have judged the existence of these entities. Even in the case of indefinite articles, 'A stone hit a car ...', while here no objects are presupposed, the relevant *types* are presupposed.

As a broader objection, Hacker writes

There are no straightforward rules for translating ordinary event-recording sentences into the canonical notation in advance of displaying and analysing their logical structure, not in the forms of the predicate calculus, but in terms of the verbs (and their specific meanings), the qualifying adverbs (and their specific significance, and hence effect upon the overall meaning of the expression or expressions they qualify), the application of the nominalizing operation to different types of adverbially qualified verbs, etc. (1982, pp. 485–6)

So when B is said 'to have wisely apologized', we cannot merely characterize this as an event which is an act of apology, whose agent is B and which was a wise event. Instead, we must look into the specificity of the meaning of the verb 'to apologize' to bring to light what we mean by someone being wise to do this. Hacker concludes

> Failure of Davidson's programme does not reveal something mysterious, let alone awry, with our ordinary forms of discourse, but the poverty and narrowness of the predicate calculus. (1982, p. 487)

But what if, instead, we look to represent event statements as types, preferably ones with a rich enough structure to allow appropriate inferences to be made? Well, linguists have taken to heart the proposal made by the philosopher Vendler that the basic category of events is composed of four subcategories: *activities, achievements, accomplishments* and *states*. Without entering into a close discussion of this proposal, the idea here is that we have

1. **Goalless activities**, such as playing in a park, or repeatedly jumping;
2. **Achievements** marking moments of reaching some goal, such as arriving in Beijing, or closing the deal;
3. **Accomplishments** include the activity leading to the culmination of a goal, such as running three miles, or emptying one's plate;
4. **States** involve no activity, such as to be square, or to feel love for something.

Since Vendler, extra details have been added, such as subdivision of achievements to account for differences in implicatures depending on whether or not the achievement is realized by an agent or not. For instance, in the case of the second sentence of the following pair but not the first, we may ask whether the event was deliberate or not:

- The lava flow reached the city wall.
- Petra lifted up her arm.

Furthermore, to Vendler's list, Bach (1986) added 'momentaneous' non-goal-directed happenings, such as 'Jane sneezed' or 'the horse stumbled'.

Linguists have realized that there is no fixed association between verb and event type. For instance, we often use an accomplishment and the imperfect tense to express an activity: *I am climbing Mount Snowdon. I am baking a cake.* This does not have as an implication that the achievement will come to pass. I may have to stop my ascent, or a power cut may interrupt my baking. So it is easy to think of cases where a verb appears in an accomplishment and an activity, such as 'climb' in the following sentences:

- Kit climbed the mountain peak.
- Kit was climbing the mountain, but didn't make the peak.

The saliance of this accomplishment–activity distinction is marked linguistically by the use of 'in' and 'for' when time intervals are assigned:

- Kit climbed the mountain peak *in* six hours.
- Kit was climbing the mountain *for* six hours, but didn't make the peak.

Moreover, sometimes work has to be done by the reader or listener to force or 'coerce' an interpretation of the use of a verb to a suitable event type. The *sneezed* in 'He sneezed as she entered the room' may be a momentaneous happening, but if I read 'He was sneezing while the choir finished the oratorio', to make sense of the use of the imperfect I need to blow up this point-like event by iteration to an activity in progress. Here 'sneezing' means having a bout of sneezes. This is represented in the lowermost arrow of Figure 2.1 below, from Moens and Steedman (1988):

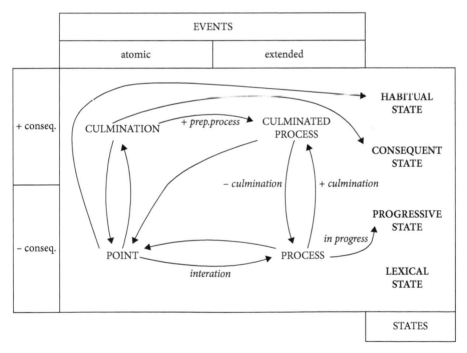

**Figure 2.1**

To interpret this diagram in Vendler's terms, we must make the following identifications: Process = Activity; Culminated Process = Achievement; Culmination = Accomplishment; Point = Momentaneous Happening. Arrows in the diagram are regular assignments, so that, for instance, the map from Culmination to Consequent State sends an achievement to the state of having achieved it: for example, *reach Cairo* to *having reached Cairo*. We can say that the *iteration* and the *consequent state* arrows, as well as the others, are mappings between types. My travelling to Cairo last year is an element of the type of such culminated processes, *Travel to Cairo*. This latter type is an element of the type *Culminated process*. *Travel to Cairo* is mapped by the arrow '—culmination' to *Proceeding towards Cairo*.

Moens and Steedman proposed that we see these event categories in relation to what they term an *event nucleus*. Typical basic events are composed of three components: a preparatory phase involving some activity, culminating in some accomplishment, which results in a change of state for some period. For instance, I reach out to a switch and flick it on, thereby lighting the room. These three components comprise an *event nucleus*.

- Event nucleus = 'an association of a goal event, or *culmination*, with a *preparatory process* by which it is accomplished, and a *consequent state*, which ensues.' (Moens and Steedman 1988, p. 15)

This tallies with our discussion in the first section of this chapter of the need to perceive a firm temporal boundary to a process or activity to count it as constituting an event. Its culmination in a change of state provides a first-rate way to identify an event.

In the context of this chapter, we might say that this event nucleus takes on the form of an iterated dependent sum, like the context associated to a very short story.

$$Event\ nucleus := \sum_{\substack{x:\,Activity \\ y:\,Achievement \\ z:\,State}} (Culminate(x,y)\,\&\,Consequent(y,z))$$

Since

$$Accomplishment := \sum_{\substack{x:\,Activity \\ y:\,Achievement}} Culminate(x,y)$$

we can see that

$$Event\ nucleus := \sum_{\substack{w:\,Accomplishment \\ z:\,State}} Consequent(q(w),z),$$

where $q$ projects onto the second component of an accomplishment, namely, the achievement.

'*The stone hit the car*' is a happening. If it were part of an intended action '*John hit the car with a stone*', this would be an achievement representing the successful culmination of an activity. We therefore expect there to be a description of this activity giving rise to it, such as John planning his attack, selecting a stone and launching it. We also expect effects, consequent changes of state, such as a dent made on the car. If we learn that it is true that this event happened, then we infer that the stone was moving relative to the car right before the moment of impact, that they came together and this produced a noise.

Evidence that we operate with the event nucleus schema in our minds as we look to understand the description of events comes from our rejection of some sentences as ungrammatical, and also from the effort we experience in making sense of a sentence which is difficult to parse. Take the first of these cases. We find it awkward or even wrong to say

- Paul broke the plates off the table,

while we happily say

- Meg brushed the crumbs off the table.

Linguists explain this difference by observing that the verb 'broke' already depicts a change of state, and so there is no place to add 'off the table'. On the other hand, 'brushed' merely depicts an activity. Completion of the event nucleus requires a resulting change of state.

As for difficulty in parsing, consider

- It took me two days to learn to play the Minute Waltz in sixty seconds for more than an hour.

'For' and 'in' are used to mark periods of time in, respectively, activities and accomplishments. 'Play the Minute Waltz in sixty seconds' is therefore an accomplishment, and yet it takes part in an activity, something lasting *for* more than an hour. This can only happen by iteration transforming an accomplishment, reduced to a point, into an activity. Finally, there is the accomplishment of learning to be able to perform such a feat. There is a preparatory process lasting two days which culminates in the acquisition of this ability.

Our desire to complete an event nucleus makes sense of requests for explanation. Along with deductive-nomological, inductive-statistical and causal explanations, when it comes to animals we often seek a teleological explanation, a sought-after change of state brought about by some activity. *'Why is that animal digging?'* 'It's burrowing. It wants to prepare a place to lay its eggs safely.' We look to place an activity with the setting of an event nucleus.

Similarly with humans and designed instruments:

- *Why are you flicking that switch?* I want to turn on the light. I believe this switch is for the light.
- *Why is the thermostat clicking on now?* It's to start up the boiler. It's designed to do so when the temperature drops.

Upstream, there will be some disposition, evolved or designed, or belief-desire bringing about the behaviour.

When someone poses a question of the form '*Why b?*' where $b$ : *Activity*, it is likely that they are wanting to know if $b$ is preparatory for some achievement so as to bring about some change of state. So in response, I may offer an element of *Achievement, Accomplishment* or *State. Why are you going up into the attic? To fetch my tennis racket. I want to have my tennis racket.* I may also be asked to fill in the first components of an event nucleus. *How did you get hold of a tennis racket? I went into the attic to fetch one.*

Such explanations rely on plausibly reliable connections between their parts, sometimes in the form of rules:

> Suppose you find yourself in a situation of a given type $S$; and suppose you want to obtain a result of a given type $R$, and there is a rule that in a situation of type $S$ the way to get a result of type $R$ is to do action of type $A$. (Collingwood 1939, p. 103)

Collingwood goes on to point out the limitations of this framework for action: that situations may be too specific to have rules applicable, and so on. This points to a distinction between particular activities culminating in particular achievements, and activity subtypes regularly culminating in achievement subtypes: for example, dropping an egg from high above a hard floor generally culminates in a broken egg. For the mathematical physicist, Hermann Weyl, it is at this level that we can speak of causal relations:

> The phenomena must be brought under the heading of concepts; they must be united into classes determined by typical characteristics. Thus the causal judgment, 'When I put my hand in the fire I burn myself', concerns a typical performance described by the words 'to put one's hand in the fire', not an individual act in which the motion of the hand and that of the flames is determined in the minutest detail. The causal relation therefore does not exist between events but between types of events. First of all–and this point does not seem to have been sufficiently emphasized by Hume–*generally valid relations must be isolated by decomposing the one existing world into simple, always recurring elements.* The formula '*dissecare naturam* [to dissect nature]' was already set up by Bacon. (Weyl 1932, p. 56)

Of course, not all activities are directed to goals, and some activities are directed to multiple goals. The type

$$\sum_{y:\,Achievement} Culminate(b, y)$$

for some activity, $b$, is a set which need not be a singleton, and may also be empty.

Events are clearly closely related to time. We expect a particular activity to take place in an interval of time, and any culmination to happen at an instant, or at least within a relatively short time interval. 'The orchestra completed the full performance of Beethoven's ninth at 10 pm.' We would reasonably expect the orchestra to be playing at 9.30 pm, since there is mention here of an achievement. Some act has been accomplished, and we may know the preparatory process takes a little more than an hour. But perhaps there was a bomb scare, and after the first three movements starting at 8 pm, the audience were evacuated around 8.45 pm and only returned at 9.35 pm. The orchestra nobly decided to complete the symphony, despite the continuing threat. Should these be considerations with which logic is to concern itself? Well, we may robustly infer that the orchestra was playing Beethoven's ninth some time before 10 pm.

Time can certainly be used to distinguish events. The plague of 1348 cannot be the plague of 1665, even if the infectious agent is the same. But of course there's more to the identity of events. Is my flipping the switch the same event as my turning on the light? For Davidson,

*Yes*, for Kim, *No*. With my type-theoretic reading in place, we can say that both are right in a sense. *Flipping the switch* is an accomplishment. As for *turning on the light*, one natural reading is as an achievement resulting in a change of state. On the other hand, it might be taken as a full event nucleus with accompanying accomplishment. Certainly, if a begrudging person dawdles on their way to the switch, you might say exasperatedly '*How long does it take to turn on the light?*' Either way, strictly speaking, Kim is right, putting us also on the side of Goldman (2007, pp. 466–7) who counts himself, like Kim, as a *multiplier*, unlike the *unifiers*, such as Davidson.

A more protracted description separates out the parts of the event '*Oliver moves his finger, flips a switch, turns on a light, and alerts a prowler.*' For Goldman, the unifier sees one action, where as the multiplier sees four distinct actions. From our perspective, we have a clear event nucleus, and then an unintended consequence. As mentioned in the first section, we should only count within a type, having individuation rules established there, whereas Davidson is taking events as a basic type and then individuating according to their causal relations.

Goldman appears to be a multiplier for something approaching type-theoretic reasons:

> if action *a* is a token of action-type *A* and action *b* is a token of action-type *B*, then *a* is not identical to *b*, even if *a* and *b* are performed by the same agent at the same time. (Goldman 2007, p. 472)

But the unifier's opposite point of view can also be given a type-theoretic gloss. It certainly seems most reasonable to be a unifier in cases where the difference concerns fine-grained levels of descriptions of some activity or accomplishment, as made apparent by the presence or absence of an adverb. So unifiers and multipliers disagree as to whether, say, Caesar dying violently is a different event from his dying. Davidson's thought is that this adverbial modification resembles the adjectival modification of a noun. In the latter case, we typically take an apple before us as the same thing as the red apple before us. Strictly, in *intrinsic* varieties of type theory, such as HoTT, since a judgement as to something being a red apple requires more than a judgement as to something being an apple, then these are simply different types. Indeed, the former type is defined as a dependent sum $\sum_{x:Apple} Red(x)$. So similarly, violent deaths are formed as a dependent sum on the type of deaths, a subtype of momentaneous events. But we should note there exist *extrinsic* varieties of type theory which do allow an element to belong to more than one type. We shall return to this issue in the following chapter.

We have seen, then, that a number of philosophers instinctively take a type-theoretic turn, without explicitly embracing the calculus fully. Given the intricate structure operating behind our ways of seeing and speaking about the world, if a formalism is to have any chance of capturing this structure, it will need to be at least as sophisticated as a dependent type theory. But through this chapter we have also seen the need for a spatial and a temporal dimension to be included. These are very much to the fore in further of Hacker's objections to the treatment of events by predicate logic. He notes that we rely greatly on contextually intricate adverbial constructions, as in '*A was often drunk on New Year's Eve*'. This clearly doesn't mean that '*A was often drunk*', since *A* may refrain from consuming alcohol for the rest of the year. To unpack the former claim, we would say that there exist multiple events such that they involve *A* drinking alcohol in a limited time period and achieving inebriation. Relative to all final days of the year, inebriation was achieved frequently, but we have no information for other days. We clearly need to be able to represent rates of states within time slots, requiring, perhaps, a form of temporal logic. For a type-theoretic approach to temporal logic, the reader will have to wait until Chap. 4. Continuing here briefly, in a similar vein Hacker points out that the following analysis of a further case will not do:

> 'A scarcely moved' as 'there was a moving by A and it was scarce', or 'it was done in a small way'. (Hacker 1982, p. 485)

Along with time, we need here some underlying space, and positions within that space depending on the passage of time through some interval. Evidently, these are some of the ingredients which type theory will need to represent when it comes to physics.

Finally, returning to 'B wisely apologized', we recognize an achievement involving an agent, *B*, where it is presupposed that there is some (here unnamed) recipient (or recipients) of the apology. Such an achievement type comes also with a presupposition that some prior offence by *B* has been recognized by the party to whom the apology is addressed, and that the apology will change the existing state of offence with likely beneficial results, such as the removal of the threat of revenge. Now, perhaps one might suspect that any inference involved here is very much *material*, highly dependent on the specific meaning of the words involved, but, as with the example from Chap. 1 of the red ball that is therefore coloured, it would be worth exploring whether we can use type theory to make sense of the structural features of even such sophisticated intentional actions as apologizing.

## 2.7 Revisiting the Philosophical Literature

Before we move on to more mathematical waters in Chap. 3 and introduce the 'homotopy' element of homotopy type theory, let me end this chapter with a few comments about the future of dependent type theory in philosophy. I can with confidence predict that it would be immensely profitable to run through large swathes of the philosophy of language and metaphysics literature with dependent type theory in mind. Places to begin are not hard to find. Experience suggests that in a great many places thinkers are striving to manage without the proper resources of a type theory. Take, for example, the following claim:

> It is now a near-consensus view in both philosophical and psychological work on emotion that emotions are intentional in the phenomenological sense—in other words, emotions possess a directedness towards a situation *as* a situation of a certain class, which is also defined by a corresponding meaning. (Keeping 2014, p. 243)

Now, one needs a calculus which can take a kind of thing, here a *situation*, and form a class of that kind, so 'a situation of a certain class'.

Type theory offers the advantage here in that in the world of types there are no naked particulars. Or, to put this thesis in Aristotelian terms, as taken up by John McDowell, every

'This' is a 'This such'. Something being declared to be of a type means that we may already know much about it. To give the type, conditions need to be set as to the 'introduction' of terms of a type, what we can do with them, that is, their 'elimination' rules, and their identity criteria. As observed earlier, this is very close to the entitlements and commitments of Brandom's inferentialism, and I predict that his work would offer rich pickings for the type theorist.[14]

So, in type theory we find two slogans apply:

1. No entity without type.
2. No type without identity.

In our type theory we will only ask of two terms whether they are equal if they belong to the same type.[15] And remember this isn't something to be determined—terms always come typed. We pose the question of the identity of terms by construction of a type of identities:

$$A : Type, a, b : A \vdash Id_{A(a,b)} : Type$$

I can't even pose the question of the identity of two terms of different types.

In *Sameness and Substance*, Wiggins (1980) argues for the 'Thesis of the Sortal Dependency of Individuation' and against Geach's assertions concerning the 'Relativity of Identity'. The latter claims that it is possible for two individuals that they be the same according to one sortal but different according to another. Against this relative identity, Wiggins looks to demonstrate principle $W$:

$$W : (\exists f)(a \underset{f}{=} b) \supset ((g)(g(a) \supset a \underset{g}{=} b)).$$

If two individuals of the same sortal are the same, then they're the same under any other sortal to which they may belong.

In our type theory, an element strictly belongs to one type alone, so we cannot speak precisely in Wiggins's terms. We might have a base type $A$, and then dependent propositions, $x : A$, $F(x)$ and $G(x)$. By $(a \underset{f}{=} b)$ we might mean the identity type of two elements in the dependent sum, $a =_{\sum_{x:A} F(x)} b$, is inhabited. Then via projection to the $A$-component, we would have that $p(a) =_A p(b)$ is inhabited. Obviously, then Wiggins's condition would hold. If $G(p(a))$ is true, then of course so is $G(p(b))$.[16]

Alternatively, the dependent sums are taken as subtypes of $A$, where we typically consider them as 'the $A$ which are $F$' and 'the $A$ which are $G$', and speak of their elements as though

---

[14] See Paul Redding (2007) on why neo-Hegelians might want to return to Hegel in his use of Aristotle's term *logic*.

[15] But see the discussion of the difference between intrinsic and extrinsic forms of typing towards the end of §3.2.1 in the following chapter.

[16] In fact, the type-theoretic account is more general. It may matter *how* $a$ and $b$ are the same. When we have $m : a =_A b$, then for any $A$-dependent type, $P$, for $c : P(a)$ we have $m^*(c) : P(b)$.

they are simply elements of $A$. So then we might understand by $(a = b)$ that $a, b : A$ and
$f$
that $a =_A b$ and $F(a)$ and $F(b)$ are true. But then of course, when $G(a)$ is true, we also have
$G(b)$ is true, and so $(a = b)$. In sum, then, if Wiggins's considerations are to be taken up by a
$g$
dependent type theory, they appear to be rather trivially the case. So why the controversy?

Let us then consider a counterexample which Wiggins discusses, one apparently proposed by Geach and Anscombe (Wiggins 1980, p. 41n35), of the road between Athens and
Thebes. Let $x$ be the road from Athens to Thebes, and $y$ be the road from Thebes to Athens.
Then $x$ and $y$ are the same as roads, and yet $x$ may be uphill and $y$ downhill. $x$ certainly
doesn't equal $y$ as uphill roads, since $y$ isn't even an example. So it seems we might have
a counterexample to Wiggins's principle $W$.

Wiggins points out that to speak of the road from Athens to Thebes being uphill, we
are really speaking about a journey along a road rather than a road *simpliciter*, and, as such,
these two journeys are not the same. From our perspective, he is right—there are two types
involved, albeit related ones, namely, the type of roads between two towns and the type
of directed roads between towns. Of course there are mappings between these types: one
may forget the direction, and on the other hand one may take a road to the two associated
directed roads. We might then write $U : Directed\ Road \rightarrow Road$. Then the identity of $U(a)$
and $U(b)$ in *Road* does not entail identity of $a$ and $b$ in *Directed Road*. *Uphill* and *downhill*
are not predicates for the type *Road*.

We are prone to introduce an entity, then refer to it by 'it', allowing us to ask 'Is *it* the same
as something else?' Is the road *from* A to B the same as the road *between* A and B. Without
type disciplining, it's easy to get lost here and ask improper questions of identity across types.
In cases where there is at most one element of type $A$ associated to each entity of type $B$, in
other words when we have a subtype, we can ask 'Is that red apple the same as the apple I
was given yesterday?' But there are two directed roads associated to a single road, so we do
not have a subtype.

These considerations do not depend on the entities involved having some underlying
spatial or spatial–temporal basis for their identity. I can tell a similar story for non-object-
based sortals. For any two musical works, if they are the same as pieces by Bach, then they're
the same as Baroque concertos. But I should not propose the identity of a musical work and
a particular performance.

Robert Brandom comes to similar conclusions when considering sortals (2015,
Chap. 7), since for him membership of properly distinct sortals will involve different sets of
modal behaviours. Thus

> when we appreciate the modal commitments implicit in the use of *all* empirical
> descriptive vocabulary, we see that strongly cross-sortal identity claims—those that
> link items falling under different sortal predicates with different criteria of identity and
> individuation—are *never* true. (2015, p. 27)

With type theory as our means of expression, it's not just that these claims are never true.
Rather, such identity types are not well-formed, and as such are uninhabitable.

My expectation is that any such linguistic, or, if you prefer, metaphysical, puzzle finds its solution when codified in a dependent type theory, but we must leave such investigations now to take up the quest for our full type theory. So far, we have seen the benefits of a dependent type theory, where dependent types are propositions or sets. Now we must motivate the passage to higher types.

# 3

## Homotopy Types

### 3.1 Introduction

We have seen that dependent type theory possesses many novel features with respect to other logical calculi. Its types sometimes play the role of sets, but equally they may play the role of propositions. Whereas set theory is standardly formulated as a first-order axiomatic theory whose domain ranges over sets and which is equipped with a binary membership relation, dependent type theory is given by type formation rules, and rules for the introduction and elimination of terms. Some rules, such as those for dependent sum and product, apply equally to types which are sets and types which are propositions. A form of propositional and predicate logic comes built into the type theory, arising from the application of these rules to a certain class of types, the possibly dependent *mere propositions*. But sets and mere propositions are just two kinds of type, those whose associated identity types are particularly simple, and as such comprise the lower two levels of an infinitely tall hierarchy. We look now to introduce the *homotopy* into 'homotopy type theory' and so need to gain a better grip on what is meant by this hierarchy of classes of type; in order to enable this we need a clearer understanding of identity types. We also need to learn about two other important ingredients of HoTT: the so-called *univalence axiom* and *higher inductive types*.

With these constructions in place, we will proceed through the chapter to see the power of HoTT as a foundational language for mathematics. Whether we take it as an extended logic or consider it to entail that logic is to be subsumed within mathematics, an early consideration as to how to engage philosophically with HoTT will be to revisit what thinkers took to be promising applications of previous forms of logic. One of the first applications by Russell of his 'new logic' (Russell 1905) was the analysis of definite descriptions, described by Frank Ramsey as a 'paradigm for philosophy' (Ramsey 1931, p. 263n). It seems fitting, then, to see what the *homotopy* aspect of the latest 'new logic' can bring to this question. We will find in §3.2 that a reasonable way to approach this topic casts light on mathematicians' usage of '*the*' in a generalized sense, as when they say 'the product of two groups', apparently without there being a unique way to construct such a product. We can also begin to formulate an understanding of definite description in cases where a group of symmetries operate on a collection of entities.

*Modal Homotopy Type Theory: The Prospect of a New Logic for Philosophy.* David Corfield, Oxford University Press (2020). © David Corfield. DOI: 10.1093/oso/9780198853404.001.0001

In the final section, I will apply this account of definite description to the phrase 'the structure of $A$', for a general type $A$. Awodey (2014) claims that, through the *univalence axiom*, HoTT captures what is essentially right about the structuralist position. Shulman (2017) agrees, describing it as a 'synthetic theory of structures', in the sense that nothing can be said about mathematical entities defined within it, other than structurally. We should not expect then to need to, or even be able to, construct something substantially different from $A$ when contemplating its structure. This is indeed what we find.

To the extent that our new logic captures mathematical inference adequately, we might formulate an updated logicist position in the philosophy of mathematics, one conveyed, perhaps inelegantly, by the term 'neo-neo-logicism'. Where traditional logicism looks to understand mathematical entities and results as wholly given by 'logical' definitions and inferences, the question inevitably arises of the status of the inference system as a logic. Should we, for instance, consider second-order logic to be logic, as Quine famously disputed? Since a form of set theory is contained within *our* new logic, HoTT, along with much more, that mathematics is espressible within it is not a radically reductionist position. With the addition of modalities in the following chapters, spatial concepts will also fall within the logic in a synthetic way. It appears, then, that with HoTT and variations any supposed border between logic and mathematics has become very difficult to discern. But beyond this issue of representability *in principle* lies what is to my mind a much more important matter—the *naturalness* of expression made possible by HoTT.

## 3.2  HoTT Components

### 3.2.1  Identity Types

One central choice in mathematics is the basic 'shape' of the entities under consideration. A long-established choice has been the *set*, a bag of dots which are completely distinct and yet at the same time indistinguishable. Irrespective of the way one chooses to describe such sets formally, either 'materially' or 'structurally', it is an astonishing idea that mathematics could be taken to rely on one such simple conception.

The key feature of a set for our purposes is that we only ask of it *whether* two of its elements are the same, not *how* they are the same. In other words,

- for $x, y : A$, $Id_A(x, y)$ is a proposition.

However, arising from the needs of current geometry and current physics, we find that limiting ourselves to such a basic shape is a severe restriction. Beyond sets, we need:

- *Homotopy types* or *n-groupoids*: points, reversible paths between points, reversible paths between paths, . . .

If we look to provide a definition of these entities in a standard set-theoretic context, they will appear to be rather complicated. We must specify the set of points, the set of paths between any two points, how these paths compose, identity relations between paths, and so on. What HoTT can provide is a language in which these concepts are defined intrinsically. Homotopic mathematics and gauge-theoretic physics may be directly described via homotopy types with the identity structures built in. From the new perspective of HoTT, it is now the case that when one wishes to specify that all entities in a given theory are good old-fashioned sets, this requires the added complexity of enforcing an extra condition. Let's see how.

For any two elements of a type we can ask whether or not they are the same:

- Where we have a collection $A$ and two elements $x, y : A$, we form $Id_A(x, y)$.

But then we can treat the latter itself as a collection and iterate:

- From $p, q : Id_A(x, y)$, we form $Id_{Id_A(x,y)}(p, q)$.

The axiom of uniqueness of identity proofs insists that any two proofs of the sameness of entities are themselves the same. In HoTT we reject the axiom that claims this to be the case, or in other words we don't insist that the type $Id_{Id_A(x,y)}(p, q)$ is inhabited.

Now, of course, as a type theory, rules must be given to determine the nature of identity types. So there is a type *formation* rule which allows us to form a new type composed of identities between any two given elements. As we have seen, when in some context we have formed a type, $\Gamma \vdash A : Type$, then we may write:

$$\Gamma, a : A, b : A \vdash Id_A(a, b) : Type.$$

I will just give a sense of the associated *introduction* and *elimination* rules here. Identity types are a form of higher inductive type, discussed below, and so associated rules just follow on from general principles for such types. But it is useful to have a sense of what this construction produces in this particular case. First, one needs to specify which canonical terms of the types are to be introduced. Surprisingly, all that are specified are the most basic self-identities, so for any element of a type there is a designated witness to the element's self-identity:

$$\Gamma, a : A \vdash refl_A(a) : Id_A(a, a).$$

One comparison of this form of *inductive* specification would be with the construction of the type of natural numbers

$$\overline{\mathbb{N} : Type}$$

where term introduction is specified:

$$\overline{0 : \mathbb{N}}$$

and

$$\frac{n : \mathbb{N}}{sn : \mathbb{N}}.$$

However, where in a sense the natural numbers are constituted by 0 and its successors, we expect the possibility of elements in types $Id_A(a, b)$ that are not of the form $refl_A(a)$.

The following situation may help some readers, otherwise it can be ignored. Given two groups, $G$ and $H$, you can form a new one, called their coproduct, constituted by finite alternating strings of elements of $G$ and $H$. This coproduct group is being specified by its having as elements all elements of $G$ and all elements of $H$, and then the fact that it is a group dictates what the other elements must be, how they compose, and so on. Similarly, in the case of identity types, any elements and their identities arise from the specified $refl_A(a)$ elements simply by virtue of its being a homotopy type.

Then we need to know how to *eliminate* terms. The full formulation for identity types will perhaps strike the casual reader as quite complicated. In essence, the rule functions like recursion for the natural numbers. Know what to do in a situation parameterized by 0 and how to transform this under application of successor, and you know what happens for the whole type of natural numbers. Again, in the group coproduct case, to specify a map from this group to another group, all one needs to do is specify where the strings of length 1 are sent, that is, the $g$ of $G$ and the $h$ of $H$. So, similarly, if we know what happens in a situation concerning the $refl_A$ elements, then we know what happens for all the identity types $Id_A(a, b)$ and their elements by the elimination rule which *transports* along these elements.

### 3.2.2 *The Type Hierarchy*

So, all propositions and all sets are types, they are just types of different kinds. What distinguishes these kinds is their position on a hierarchy corresponding to what is called the *truncatedness* or *homotopy level* of the type. These levels are defined recursively in terms of the type's corresponding identity types. As we have just seen, no restrictions are placed on such identity types, in the sense that it is not required to be a proposition whether two elements are equal. In other words, we may have two elements, $p$ and $q$, of $Id_A(a, b)$ which are not equal. The elements $a$ and $b$ may be equal, but in different ways. Indeed, we can form a further type, $Id_{Id_A(a,b)}(p, q)$, and then iterate this process.

Informally, mere propositions are taken to be types for which any two terms are equal. There may be no such terms, in which case the proposition is false, but if there is a term, in which case it is true, then there is only one term.[1] Similarly a type is a *set* if its corresponding identity types are mere propositions, so that an answer to whether two terms are the same may be 'yes' or 'no', but there is no sense in which they are the same in a number of different ways. A type is a *groupoid* if its identity types are sets, a *2-groupoid* if its identity types are groupoids, and so on. Note that this hierarchy of levels is cumulative, in the sense that a type which is a mere proposition ($-1$-type) is also a set (a 0-type), a groupoid (1-type), and so on. Indeed, any $m$-type is also an $n$-type for $n \geq m$.

The expression '$n$-groupoid' comes from algebraic topology. They arise, for example, when considering a topological space, a collection of points in the space, paths between pairs of points, paths between paths with the same endpoints, and so on. But they may also be

---

[1] We encountered this concept already in Chap. 1 when discussing 'bracket types' to treat the inference from *is red* to *is coloured*.

treated 'algebraically' as corresponding to a special kind of $n$-category, where the morphisms, morphisms between morphisms, and so on, are all invertible up to some weak equivalence (see Corfield 2003, Chaps 9 and 10). A useful illustration of the need for higher groupoids comes from attempts to grasp the topology of the 2-sphere. Consider a point, $N$, at the North Pole and another, $S$, at the South Pole. To travel from $N$ to $S$, I can pass down any meridian. I could also travel down to the equator, do a full loop of this westwards and then proceed to $S$. All these paths are equivalent in the sense that each may be continuously transformed into any other. It seems, then, that something is missing in that we are detecting no difference in the behaviour of the surface of a sphere compared to that of the plane. There should be some way to distinguish them, and indeed there is. Consider a path down the Greenwich Meridian, $g$, and another path down the International Date Line, $i$, at the opposite side of the world. Now which paths are there from $g$ to $i$? Well $g$ can be swept eastwards or westwards to $i$, and these sweepings are not deformable to each other. I can also sweep $g$ a number of times about the world. In fact, the distinct ways of doing so are isomorphic to the integers, $\mathbb{Z}$. But the story doesn't end there. The 2-sphere is surprisingly complicated and indeed has nontrivial higher homotopy, detecting paths between paths between paths, and so on, in all degrees over 1.

Because the level measuring the varieties of non-deformable path was tagged with '1' by the algebraic topologists, paths between paths with '2', and so on, the level below, which in topological terms collects the set of connected components of a space, is assigned 0. Thus sets are seen as 0-groupoids. Naturally, mere propositions appear as $(-1)$-groupoids. We can even take one further step to a $(-2)$-groupoid, which is just a contractible space.

| | |
|---|---|
| $\cdots$ | $\cdots$ |
| 2 | 2-groupoid |
| 1 | groupoid |
| 0 | set |
| $-1$ | mere proposition |
| $-2$ | contractible space |

Numbering from $-2$ upwards may not seem optimal in retrospect, so some people will sometimes add 2 to the level to begin from 0.

Let us now see how to express the homotopy levels ideas formally. So we define for any type, $A$, the type

$$isContr(A) :\equiv \sum_{x:A} \prod_{y:A} Id_A(x,y).$$

For this to be inhabited, we must produce a specific element of $A$ and then a way of proving that any other element is equal to this one. This must happen in a 'continuous' way that respects any identities in $A$.

Then for any type, $P$, we can define what it is for it to be a (mere) proposition:

$$isProp(P) :\equiv \prod_{x,y:P} isContr(Id_P(x,y)).$$

This requires of any two elements of $P$ that their identity type is contractible, that is, that there is one way in which they are the same. It can be shown that an equivalent definition of *isProp* is given by:

$$isProp(P) :\equiv \prod_{x,y:P} Id_P(x,y).$$

If we have an element of this type here, that is, a function, then it provides an identification between $x$ and $y$, and again does so *continuously*, which implies not only that any two elements of $P$ are equal, but also that $P$ contains no higher homotopy.

The definition for a type being a set is now defined in terms of its being the case that the identity type for any two elements is a proposition:

$$isSet(S) :\equiv \prod_{x,y:S} isProp(Id_S(x,y)),$$

and so on by iteration.

Propositions form a subtype of any type of types, $\mathcal{U}$, and this is designated

$$Prop :\equiv \sum_{X:\mathcal{U}} isProp(X).$$

Any type may now be 'truncated' to form a proposition

$$A : Type \vdash ||A|| : Prop$$

with $a : A \vdash |a| : ||A||$. This marks the passage from, say, a type *Cat* to its *bracket type*, the proposition *There is some cat*, where all we care about is that there be a cat, without any interest in how many or their identities. If Tibbles and Kitty are cats, then $|Tibbles|$ and $|Kitty|$ are equal as evidence for the proposition *There is some cat*. The passage from occasions when Kim drank coffee yesterday to the proposition *Kim drank coffee yesterday* would be another example. Now, through the truncation, Kim's morning latte and her afternoon espresso become identified as embodying the truth of her coffee-drinking yesterday.

The notion that propositions, sets and other $n$-groupoids share a common nature by being types, and so are subject to the same rules, takes some getting used to. Philosophers have generally taken it for granted that the application of mathematics to the world requires a response from them in a way that the application of logic does not. So *indispensability* arguments propose that the successful use of, say, some simple set theory or even basic arithmetic should force us to commit ourselves to the existence of abstract sets or numbers, unless we can wriggle out of these commitments, say, by rewriting our applied mathematics so as not to mention numbers (see Field 1980), as with

$$\exists x \exists y (Ax \wedge Ay \wedge x \neq y \wedge \forall z(Az \rightarrow (z = x) \vee (z = y))),$$

to state that there are precisely two $A$s. From the perspective of HoTT, however, applied arithmetic should incur the same price as applied logic. *Logical inferentialists* have made the case that connectives such as 'and' and 'implies' take their meaning from the introduction and elimination rules they obey. If we could make a convincing case to carry this inferentialism over to the whole of HoTT, and as a type theory formulated according to introduction and elimination rules this seems eminently plausible, then we might take ourselves to have avoided making reference to abstract mathematical entities. Let us see how we might begin to make such a case for a simple arithmetic situation.

Since all $n$-types are to be treated evenly, along with *propositional* truncation to the bracket type, we may also truncate to any level in the hierarchy: $||A||_n$ is the truncation of $A$ to an $n$-type and $|a|_n$ is the image of an element, $a$, of $A$ in the truncation, where the subscript is generally omitted when $n = -1$. Each of these truncation operations is *idempotent* in the sense that applying the construction twice is equal to applying it once. These will give us examples of 'modalities' to be taken up in the following chapter. For example, we can form the set $Card :\equiv ||Set||_0$ of cardinalities (§10.2, UFP 2014). Since we can form by induction in HoTT a family of types

$$n : \mathbb{N} \vdash Fin(n) : Type$$

(by iterative type summation of the unit type $\mathbf{1}$ to the empty type $\mathbf{0}$) which are standard finite sets of cardinality $n$, there will be an inclusion from $\mathbb{N}$ into $Card$. Saying 'There are $n$ As' for a type $A$ is to say that $A$ is a set, and that $|A|_0 = 2$ is true, or equivalently that $||id_{Type}(A, Fin(2))||$ is true, meaning that there exists an isomorphism between $A$ and the standard two-element set. A type-theoretic version of Fieldian rewriting in terms of dependent sum and product to express the existence of precisely two elements in a type is possible, but (since provably equivalent to the above formulations) unnecessary. As to whether 'ontological debt' has been incurred, this will rest on what one makes of the type formation rules.

Since the standard cardinality applies to those types which are sets, this points to the fact that we will have to be more subtle when thinking about extending such a notion of size to types which are groupoids and higher groupoids. But there is such an equivalence-invariant construction. This is, for example, in the case of a (1-)groupoid, the sum of a contribution from each of its connected components equal to the reciprocal of the number of arrows at any object. This construction has a point beyond mere mathematical interest. Cardinalities of higher groupoids appear as the partition function in extended topological field theories in physics (see, for example, Freed and Teleman 2018, §9.1).

### 3.2.3 The Univalence Axiom

Through dependent types we see the need for another central ingredient of HoTT, namely, its universes, or types of types. A very important idea in mathematics is the notion of a *moduli space* or a *classifying space*. I have some space or set, $X$, and I want to know about all the ways a certain structure may be placed on it. Often I can find a mathematical object, $A$, such that maps to that object correspond to structures of that kind on $X$. $A$ is often called a *moduli space* or a *classifying* space, modulating or classifying the variety of such structures. It is also called a *representing* object. This plethora of names is a sign of the centrality of the concept.

An easy example concerns equipping a set with a designated subset, or equivalently a property. The moduli space for such equippings is a two-element set. One can think of the designated subset fibred above the given set with fibres of either 0 or 1 elements. To illustrate this, if I ask a group of people to raise their right hand when their birth year is even, we can see this fibring realized. Above the even-yeared, an arm extends, topped by the point of their hand; above the odd-yeared, there is nothing.

$$
\begin{array}{c}
\textit{Even-yeared people} \\
\downarrow \\
\textit{People} \quad\quad \rightarrow \quad \mathbf{2}
\end{array}
$$

Each different subset of *People* can be picked out by a different map to **2**.

More sophisticated examples include classifying spaces in algebraic topology and moduli spaces in algebraic geometry. For the former, we might have a space, $B(U(1))$, such that different maps to it from a space $X$ correspond to different $U(1)$-principal bundles, where each fibre above a point of $X$ is a copy of the complex circle $U(1)$. This construction is employed when dealing with electromagnetic gauge theory. We also find moduli spaces throughout mathematics: for example, for curves, varieties and schemes in algebraic geometry.

The case above of the two-element set as classifier of subsets is an example of a more general category-theoretic construction. The category of sets is a topos, the kind of category in which it is possible to reason largely as though dealing with sets, functions, predicates, relations, and so on, though restricting oneself to constructive inference. A vital ingredient to what makes this possible is the presence of a *subobject classifier*, sometimes called an *object of truth values*, usually denoted $\Omega$. Such an object comes endowed with a structure, in general an *internal Heyting algebra*. This means that the maps from an object, $A$, to $\Omega$ inherit this structure.[2] In other words, the subobjects of an object $A$ form a kind of lattice, and in the case of a *Boolean* topos, such as the category of sets, this lattice of subsets is a Boolean algebra. Now when we move up to an $(\infty, 1)$-topos, we require a special object which not only classifies subobjects of a given object, but also controls the way other objects map to it. To see this, let's now shift to the corresponding type-theoretic perspective, where we have already glanced at such an entity whenever we have written $A : Type$ or $A : \mathcal{U}$.

It is clear that for something to play the role of the classifier of dependent types over a given type, $A$, we need a type for which different maps to it out of $A$ correspond to different types depending on $A$. Since for a dependent type, $B$, as $x$ varies over $A$, $B(x)$ is a type, it would seem that we need a type of types, $x : A \vdash B(x) : Type$, that is, an 'object classifier'. Alarm bells should be ringing on hearing about a type of types, but all is kept under control by the provision of a nested hierarchy of type *universes*, $Type_i$. As the HoTT book observes, rare cases where one needs to speak of multiple universes are readily treated (UFP 2014, §1.3). Other than here now to point out that a type of types cannot be an element of itself, I will not need to deal further with this issue, and will just use '$Type$' or '$\mathcal{U}$' when I need to mention a universe. So, dependent types are presented by maps, $B : A \rightarrow Type$, and can be pictured as a downward arrow projecting onto $A$, for example:

$$
\begin{array}{c}
\textit{Players} \\
\downarrow \\
\textit{Teams} \quad \rightarrow \quad \textit{Type}
\end{array}
$$

---

[2] Strictly speaking, the *internal hom* $[A, \Omega]$.

This universe object sitting in an $(\infty, 1)$-topos is endowed with a great deal of internal stucture which allows it to dictate the logic of its ambient setting.

Now, two important constructions, both mathematically and in physics, are *the total space* and *the space of sections* of a fibre bundle. The total space brings together all of the elements in all of the fibres over $A$, so giving us back our dependent sum $\sum_{x:A} B(x)$. On the other hand, sections of a fibration are maps, $\sigma$, from $A$ to the total space picking out for each element, $x$, of $A$ an element of the fibre, $B(x)$; in other words, elements of our dependent product $\prod_{x:A} B(x)$. In terms of physics, a section of a principal bundle over space-time is a *field*. We represent such a section by an arrow upwards, so that post-composition with the projection down is the identity

$$
\begin{array}{c}
Players \\
\downarrow\uparrow \\
Teams \quad \rightarrow \quad Type.
\end{array}
$$

For instance, the team to which the captain of $t$ belongs is $t$ itself.

In the case of a dependent proposition, such as $x : Person \vdash Even\text{-}yeared(x) : Prop$, represented by

$$
\begin{array}{c}
Even\text{-}yeared\ person \\
\downarrow\uparrow \\
Person \qquad \rightarrow \quad Prop,
\end{array}
$$

we find that the dependent sum is the type of even-yeared people, while the dependent product will only be inhabited if everyone is even-yeared. Hence

$$
\forall_{x:Person} Even\text{-}yeared(x) \simeq \prod_{x:Person} Even\text{-}yeared(x).
$$

The dependent sum here will form a set, but with the truncation construction of §3.1.2 we can formulate standard existential quantification:

$$
\exists_{x:Person} Even\text{-}yeared(x) \simeq \Big|\Big| \sum_{x:Person} Even\text{-}yeared(x) \Big|\Big|.
$$

Here we only care that there be an even-yeared person, not how many or their identities.

Now, the special ingredient that makes HoTT tick is the univalence axiom. This concerns the way the universe of types, $\mathcal{U}$, behaves. If we take elements of this type to be maps $\mathbf{1} \rightarrow \mathcal{U}$, then, as a map to an object classifier, we can see these elements also correspond to types depending on $\mathbf{1}$. However, all types can be said to be so trivially dependent. The universe type, $\mathcal{U}$, is in a sense reflecting within itself the workings of all the types as a whole. The HoTT book employs the convention where the same letter is deployed to denote a type and its corresponding element of $\mathcal{U}$, and I will follow it in this respect. It is worth observing, however, that some traditions in type theory distinguish between elements of $\mathcal{U}$ and types, either by writing $\ulcorner A \urcorner$ for the element of $\mathcal{U}$ corresponding to type $A$, or by designating an element, $\mathcal{U}$, by, say, $B$, and then the corresponding type $El(B)$.

So given two types, $A$ and $B$, we have $A, B : \mathcal{U}$ and can therefore form the identity type $Id_{\mathcal{U}}(A, B)$. We should consider, then, the extent to which this type reflects the relationship between $A$ and $B$. So from these types we may construct the type of functions, $[A, B]$, and specify a subtype of this made up of functions which are *equivalences*. For a map between two sets to be an equivalence means it should be an isomorphism. This may be characterized as saying that the inverse images, or fibres, of the map are singletons. Such a map, $f : A \to B$, is one-to-one between the sets, so that for a $b$ in $B$ there is precisely one $a$ in $A$ with $f(a) = b$. But in general, for other kinds of type, we need a more subtle account. Consider a groupoid with two objects and a single arrow each way between them. This groupoid is equivalent to the trivial groupoid with a single object and its identity arrow. The inclusion of this single object into the former groupoid does not provide an isomorphism between the sets of objects, and yet it is an equivalence.

Although nothing maps directly to $Y$, the $X$ on the left-hand side is in $Y$'s *homotopy fibre* for this inclusion in the sense that there is an arrow from the target $X$ to $Y$ in the groupoid on the right. In general, then, an equivalence, $f$, between two types will be such that the homotopy fibres are all contractible. Note that the groupoid cardinality of each groupoid here is 1, an equality we should expect from their equivalence.

So then $Equiv(A, B) :\equiv \sum_{f:[A,B]} isEquiv(f)$ is a type and $Id_{\mathcal{U}}(A, B)$ is also a type. How are they related? Well, there must be a map from $Id_{\mathcal{U}}(A, B) \to Equiv(A, B)$ by the inductive definition of $Id$ types. Indeed, since there is an obvious assignment which sends $refl_A : Id_{\mathcal{U}}(A, A)$ to $Id_A : Equiv(A, A)$, we have a map given for the whole of $Id_{\mathcal{U}}(A, B)$ by the elimination rule for identity types, described in 3.1.1.

The univalence axiom says, then, that this map we have just constructed is an equivalence,

$$Id_{\mathcal{U}}(A, B) \simeq Equiv(A, B).$$

Whenever we have constructed an equivalence between $A$ and $B$, we can associate to it a term in the $Id$ type, and then use this to transfer by path induction constructions pertaining to $A$ to $B$ along this equality.

When the univalence axiom is restricted to propositions, it results in what is called *propositional extensionality*, often written as

$$P \leftrightarrow Q \simeq P = Q$$

for propositions $P$ and $Q$. This would appear to have the counterintuitive consequence that any two propositions we have shown to be true would then have to be seen as equal. The key here is to think of this as a matter of *reference* rather than *sense*. We might say that, if true, both propositions refer, as Frege thought, to *the True*. Of course, they may be conceptually very different and not be definitionally equal. Definitional equality is a different matter, corresponding more closely to Frege's *sense*.

Mathematicians are used to comparing definitions of propositions for their sense, but definitions of other things (sets, numbers, functions, and so on) for their reference. We think of '1 + 3' and '2 × 2' as the same number, where what is the same is their reference, namely 4, and not their sense. But we think of the Poincaré conjecture and Fermat's last theorem as different propositions, though what differs is only their sense now that they have been proved; their reference is the same, namely the truth value 'true'. The Poincaré conjecture and Fermat's last theorem are not *definitionally* equal, but they are propositionally equal, qua true propositions.

A related confusion arises sometimes with people imagining that isomorphic subtypes will be identified in HoTT, so that we will be forced to say, for example, that the even natural numbers are the same as the odd natural numbers. Well, considered just as bare types, $\sum_{n:N} Even(n)$ and $\sum_{n:N} Odd(n)$ *are* isomorphic sets, and so equal as types. However, with subtypes of a given type, the relevant identity is not just equivalence, but rather equivalence in the context of that type. For subtypes, this is a case of propositional extensionality in the context of that type.

$$n : N \vdash Id_{Prop}(Even(n), Odd(n)) : Prop$$

is evidently not inhabited, as even natural numbers are not odd. As expected, subsets will be equal if and only if they coincide. The fibres above the elements of a set, corresponding to a subset, are either occupied or not, and for two subsets to match, these fibres must match.

The identification of propositions and sets as just two amongst many kinds of types requires a very radical change in thinking. In analytic philosophy, propositions are singled out for special treatment, as when people talk about 'propositional attitudes'. With our type theory, however, establishing that a proposition is the case and presenting an element of a set become very similar activities. Surprise on first encountering this idea might be expressed along the following lines:

> Are you really saying that all true (mere) propositions and all singleton sets are equal? '2+2 = 4' and 'the set of even primes' are equal!

We must be very careful here. For one thing, to the extent that *the set of even primes* is constructed as a subset of the natural numbers, this cannot be compared to a proposition. It is only as an abstract set which no longer remembers that its element is called '2' that we can at least take it to be the same as any other singleton set.

To compare an abstract singleton set with an abstract true proposition, strictly we also must forget that the former is a set while the latter is a proposition. The type of sets is $(X,s) : Set_{\mathcal{U}}$, where $s$ witnesses the Set-hood of $X$. The type of propositions is $(Y,t) : Prop_{\mathcal{U}}$, where $t$ witnesses the Proposition-hood of $Y$. Before comparison can take place, we need to shed these second components, so that we can compare $X$ and $Y$ within $\mathcal{U}$. If $X$ happens to be a singleton set, then $||X||$ and $X$ are both contractible, but they're not *definitionally* equal.

A further consequence of the univalence axiom is *function extensionality* (UFP 2014, §2.9). For $f,g : \prod_{x:A} B(x)$,

$$happly : (f = g) \rightarrow \prod_{x:A} (f(x) =_{B(x)} g(x));$$

a proof of the identity of these functions can be used to establish identity of their values at each argument. Univalence then allows us to say that this is an equivalence:

$$funext : (\prod_{x:A} (f(x) =_{B(x)} g(x))) \rightarrow (f = g).$$

Again, it may seem surprising that two functions are to be treated as equal simply because we can show that they agree on their values at each argument. For example, as functions on the natural numbers, are $\lambda n.n + 1$ and $\lambda n.1 + n$ 'the same'? They are both the successor function, but this is shown in one case by simple application of the defining clauses of the *add* function (UFP 2014, §1.9), whereas it needs some work to show that this is so for the latter, involving proof by induction, and is certainly not a definitional equality. Some forms of intensional dependent type theory do not satisfy function extensionality, and may indeed treat such functions as non-identical. Here in HoTT, however, we mean that our two functions are *propositionally* equal in the type $[\mathbb{N}, \mathbb{N}]$, so that we have some $p : Id_{[\mathbb{N}, \mathbb{N}]}(\lambda n.n + 1, \lambda n.1 + n)$, generated from a proof that $(1 + n) = (n + 1)$ for all $n$.

### 3.2.4 Higher Inductive Types

One final construction we need to mention is that of *higher inductive types*. This provides a way to construct many useful types. This construction includes the ordinary inductive types, such as the natural numbers and even the empty type, as well as higher inductive types we have seen, such as identity types and type truncations.

The type *Circle*, for example, is defined so as to have a single element, *base*, and a named element in the identity type, *loop* : $Id_{Circle}(base, base)$. This marks the departure from ordinary inductive definition, since here we have a constructor not of the type being defined, but rather of an identity type, one here concerning the self-identity of a previously constructed element of *Circle*. This latter type behaves as a homotopy theorist would expect a circle to behave, for example, in terms of the type of mappings from *Circle* to itself being equivalent to the integers. Think of winding a rubber band about your finger. In standard homotopy theory, treating spaces up to continuous deformation, one might define the circle first as a subset of the real plane. The *fundamental group* of the circle will then select a point and look to distinguish between classes of equivalent (that is, continuously deformable) paths. In HoTT, *Circle* is defined as a type equivalent to this fundamental group.

Many HoTT practitioners are looking to formulate a *synthetic* homotopy theory by combining the power of the univalence axiom with higher inductive types used to define constructions there, such as the interval, the 2-sphere, suspension, mapping cylinders, pushouts, quotients of sets, $n$-truncation, localization and spectrification. This is the region of mathematics which at the moment is most amenable to a formal treatment by the type theory. With the addition of modalities (see Chap. 4) further results treating homotopy in relation to topology have been achieved, such as Brouwer's fixed-point theorem

(Shulman 2018a). This latter paper is very instructive in its way of handling a question that may already have arisen in the reader's mind.

- If HoTT is a form of *constructive* type theory, will it be possible to treat results requiring classical logic?

The answer is an emphatic 'Yes!'. Versions of the Law of Excluded Middle and of the Axiom of Choice may be added to HoTT consistently, although what may be lost in the process is a computational interpretation of the result, and the breadth of interpretation of the result in different categorical settings.

Another instance of the fertility of the capacity to define types through higher induction lies in algebra (§6.11 of UFP 2014), where it is possible, for example, to define a type which is the free group on a given set. The authors write:

> We have proven that the free group on any set exists *without* giving an explicit construction of it. Essentially all we had to do was write down the universal property that it should satisfy. In set theory, we could achieve a similar result by appealing to black boxes such as the adjoint functor theorem; type theory builds such constructions into the foundations of mathematics. (UFP 2014, p. 208)

Repeatedly through the HoTT book, evidence is given for what I termed above the *neo-neo-logist* thesis—the boundary between logic and mathematics blurs.

In sum, then, *homotopy type theory* is doubly aptly named.

- As *(homotopy type) theory*, HoTT is a synthetic theory of homotopy types or $\infty$-groupoids. It is modelled by spaces (but also by lots of other things).
- As *homotopy (type theory)*, HoTT is the internal language of $(\infty, 1)$-toposes. It is a type theory in the logical sense, and may be implemented on a computer.

Now, with the components of HoTT completed, it may surprise the reader that we turn to the humble definite article, *the*, but we shall see there are new things to learn about it.

## 3.3 Definite Description in Natural Language

We use '*the*' in a number of somewhat related ways:

- The Prime Minister of the United Kingdom is right-handed.
- The Romans invaded Britain in AD 43.
- The platypus is a nocturnal creature.

For the purposes of this section, I shall be considering its use in definite descriptions, as in the first of the examples above, where 'the' is followed by a singular noun, perhaps restricted in some way, since this seems to be the case with both Russell's 'the present King of France' and 'the structure of A'. I consider this to be the case also in standard mathematical statements such as:

- The cyclic group of order 6 has an element of order 3.

Where this last case may give the appearance of employing the kind of general 'the' used in 'the platypus' above,[3] we shall see that it is correctly taken as a case of specifying by restriction one item from a collection.

There are a range of subtleties to the use of 'the' within the singular terms of natural language; see, for example, Vendler (1967a, Chap. 2) who dwells on implicit restrictions from beyond the context of a sentence. For instance, it may be the case that an entity has been described in a previous sentence, in which case to produce the full expression we may have to insert a redundancy:

- Yesterday, I bought a car. The car is green.
- Yesterday, I bought a car. The car I bought yesterday is green.

Since our primary goal is to study mathematical usage, these subtleties of ellipsis need not detain us.

Now two related forms of such a use of 'the' present themselves:

1. 'The $A$', where $A$ is a type of a certain kind. For example, *the donkey owned by John* as an element of the type *Donkey owned by John*.
2. 'The $f(a)$', where $a : A$ and $f : B^A$, for some types $A$ and $B$ ($B^A$ being the type of functions from $A$ to $B$). For example, *the mother of Julius Caesar* as an element of the type *Woman*, 'mother of' having already been formed as belonging to the type of functions from Person to Woman, and 'Julius Caesar' as belonging to Person.

The former is properly formed when there is a unique individual of the type $A$. This may come about by forming a singleton type from an existing type, such as *Donkey owner by John* from the type *Donkey*. As we have seen, such types may be formed within the type theory using *dependent sums*. In the case of 'the donkey owned by John', if indeed it is a singleton, then there will be one pair formed of an element of the main type, here a donkey, along with a warrant that the specified condition holds; here a proof that it is owned by John.

In case (2), the fact that $f$ is a function forces the existence and uniqueness of $f(a)$, such as in the expression 'the colour of my front door', where 'colour of' is a map from some type of (monochrome) objects to the type of colours. A simple extension would allow $B$ to depend on $A$, so $f(a) : B(a)$, as in *the captain of team(a)*: *Player(a)*, for any team, $a : Team$. In either case, we might form a singleton subtype, such as 'Colour which is the colour of my front door', and in this way we can reduce to (1), 'the $A$' for some type $A$. On the

---

[3] I thank an anonymous referee for making this claim, even though I consider it incorrect.

other hand, we frequently name a function otherwise than according to the name of the target type. As Bede Rundle observes in his *Grammar in Philosophy* (Rundle 1979, p. 239), the width of the chair may be equal to the width of the door frame, but they are not the same as *widths*. Rather, they are the same as distances. Indeed, *width* is a function from a certain kind of object to the type of distances, where I can then form the proposition $Id_{Distance}(width\ of\ chair, width\ of\ door\ frame)$. Equally, in the dependent type case, I may take $captain : \prod_{a:Team} Player(a)$.

### 3.3.1 Definite Description for any Type

Although above I have been considering singleton types as though they are sets with one element, we know from the section before that types in HoTT need not be sets, but may be 'mere propositions' or 'higher groupoids' (UFP 2014, Chap. 3).

On the face of it, then, given two sets, $A$ and $B$, we might imagine that the type of products of $A$ and $B$ would form a groupoid rather than a set, there being a set of ways that two constructed products are isomorphic. To count as such a product, any such type must be equipped with a projection to $A$ and one to $B$, which satisfy certain conditions. It would appear, then, that both the obvious product composed of ordered pairs, $A \times B$, and the type of reverse pairs, $B \times A$, with the projection from second place to $A$ and from first place to $B$ would represent a product of $A$ and $B$. Using the convention of naming a type by the kind of its elements, we have $Product(A, B)$ as the type whose elements are sets which behave as a product of $A$ and $B$ should. The identity type $Id_{Product(A,B)}(A \times B, B \times A)$ might then appear to contain a non-singleton *set* of elements, in which case $Product(A, B)$ is not a set.

However, it is well known that mathematicians will say '*the* product of two sets'. Category theory has explained how to think of this case as one where, when a construction has been defined by a *universal property*, it does not matter which representative one takes as product. This is because there is a *canonical* isomorphism between any two representatives, as given by the universal construction. Indeed, for $A$ and $B$ objects in a category $\mathcal{C}$, 'the' product of $A$ and $B$ is defined as an object, $P$, with arrows (projections), $p_1 : P \to A$ and $p_2 : P \to B$, such that for any $Q$, an object of $\mathcal{C}$, equipped with maps, $f$ to $A$ and $g$ to $B$, there is a unique arrow $t : Q \to P$, such that $f = p_1 \circ t$ and $g = p_2 \circ t$. An early exercise in category theory has one demonstrate that given two such products, $P_1$ and $P_2$, there are unique arrows in each direction between them, which when composed in each order yield identity maps. This establishes that $P_1$ and $P_2$ are isomorphic, and canonically so, as the isomorphism derives from specified unique arrows.

Category theory and type theory work hand in hand here. The universal nature of the product construction applied to all homotopy types as described by the former is perfectly captured in the combination of the four rules of type formation, term introduction, term elimination and computation (UFP 2014, §1.5). Due to the conditions of the definition of 'Product', there is in fact only a single element of the identity type $Id_{Product(A,B)}$ $(A \times B, B \times A)$, namely, the map which reverses the order of the pairs. We could say that the type of products of two sets is a groupoid in which objects (namely, sets behaving as a product) are related coherently by unique morphisms, or in other words that the groupoid of products is equivalent to the trivial groupoid. Up to equivalence as a homotopy type

(UFP 2014, §2.4), such a groupoid is equivalent to a singleton set, and so *contractible* in the language of §3.1. That a contractible type be describable as a singleton set or as a trivial groupoid reflects the convention stated above that the level hierarchy is cumulative.

Recall the construction in HoTT which defines what it is for a type to possess this property (UFP 2014, §3.11):

$$X : Type \vdash isContr(X) \equiv \sum_{(x\,:\,X)} \prod_{(y\,:\,X)} Id_X(x,y) : Type.$$

To find an element of this dependent sum requires us to produce an element, say, *a*, of the type *X*, and then an element of the subsequent dependent product. To specify such an element, we are looking for a coherent (or continuous)[4] collection of identities between *a* and each element of the type. In the case of the type *Product(A, B)* for two sets, *A* and *B*, the type containing any type that acts as such a product, we have a representative, $A \times B$, and for any other representative, a canonical isomorphism, such as the switching map from $A \times B$ to $B \times A$.

One level down, consider when *X* is a set in the HoTT sense. Then we find that *X* is contractible precisely if we can find an element, and every element is equal to this one. In other words, as expected, a contractible set is a singleton. On the other hand, when *X* itself is a 'mere proposition', then contractibility amounts to *X* being inhabited, and so being true.

My proposal, then, is that we should only form the term 'the *X*' for a given type *X* once we have established that *isContr(X)* is inhabited. Of course, in the context of an assumption that *X* is contractible, we should be able to form 'the *X*' as a term depending on the type *isContr(X)*, but until we have constructed an element of *isContr(X)* we cannot form 'the *X*' in an assumption-free way. In the case of the type *Present King of France*, without the possibility of establishing unique existence since it lacks any element, Russell's term 'the present King of France' should not be introduced in a non-hypothetical way, in which case it is not available to be used to construct the proposition 'The present King of France is bald'. Rather than conjoining presuppositions into the full expression of a proposition, Martin-Löf-style type theories such as HoTT form a proposition, as any type, within a *context* (Chap. 2, §5), constructed with a valid dependency structure.[5] Conditions must be in place for constructions to be permissible.

The type theorist might say that a judgement of unique existence is *presupposed* along the lines we outlined in §2.5 above. Without this presupposition, questions concerning the possession of properties simply don't arise. This finds an early echo in Collingwood's *An Essay on Metaphysics*: 'To say that a question "does not arise" is the ordinary English way of saying that it involves a presupposition which is not in fact made.' (1940, p. 26). Indeed, as I suggested in Chap. 2, Collingwood's outlook appears to chime well with type theory. Following on from Collingwood, in his criticism of Russell's account Sir Peter

---

[4] There is a subtlety here that interested readers can read about in UFP (2014, Remark 3.11.2).

[5] Contexts appear on the left-hand side of the symbol '⊢'. $\Gamma \vdash a : A$ expresses the judgement that *a* belongs to *A* under the assumptions in $\Gamma$. Here, *A* and *a* typically depend on variables appearing in $\Gamma$. See UFP (2014, Appendix) for details.

Strawson (1950) also speaks of questions concerning the King of France not arising (ibid., p. 330), and elsewhere of the existence and distinguishability of something answering to a definite description as 'presupposed and not asserted in an utterance containing such an expression' (1964, p. 85).

Now we can formally describe the rule of what we might call *the introduction*:

$$X : Type, (x,p) : isContr(X) \vdash the(X,x,p) \equiv x : X.$$

In natural language, once we have a type, $A$, and have established the existence of a unique member, $a$, of $A$, we say merely 'the $A$' rather than tagging this term with the information $a$ and the proof that it is unique. In, say, 1780, when Louis XVI could provide the first component of an element of *isContr(Present King of France)*, then we could introduce the term 'the present King of France' in a non-hypothetical way as equal to Louis XVI. In cases in mathematics where $A$ is not necessarily a set, we may remain sensitive to the mode of construction of the $a$ appearing in the element acting as a warrant for contractibility, although we are usually less sensitive to $p$, the proof of uniqueness. Syntactically there is a difference between 'the $(A, a, p)$' and 'the $(A, a', p')$', although there is a canonical identity between these terms in $A$. Since *isContr(X)* itself is a mere proposition (UFP 2014, Lemma 3.11.4), any two of its elements are equal.

We should specify that this rule is intended for those types which are named as concepts whose instances are their elements. As we have seen in §1.4, HoTT also permits the construction of 'higher inductive types' (UFP 2014, Chap. 6) which allows, amongst other things, for the construction of types in which identities behave like the path spaces of topologically (or better, homotopically) intricate spaces. A practice has begun of naming some of these higher inductive types after the spatial properties of the type as a whole. Hence in some discussions we find *Circle*, *Interval*, *2-Sphere*, and so on. Perhaps we could say, then, that *Circle* is being used as a shorthand for 'the type which behaves like a circle', and as such is formed by 'the introduction' from the type of types that behave like a circle. This is justified, since we can show that the type of types equal to a specific higher inductive type is contractible, and so a suitable place to apply 'the introduction'.

Returning to the 'the introduction' rule above, now we see that contractibility makes sense of our application of 'the' to an apparent groupoid such as 'the product of two types $A$ and $B$'. One might think that there could be many ways to produce such a product, but the universal property defining what it means to be a product ensures that any candidate is uniquely isomorphic to any other. It is perhaps illuminating, then, to consider on this reading that, strictly speaking, for a type which is a groupoid in which every pair of elements is isomorphic, but *not canonically so*, we should *not* apply 'the' to the type. We see this in the mathematical construction of algebraically closing a field, where there is a reluctance to say '*the* algebraic closure of field $F$' although all such closures are isomorphic, since they are not uniquely so (Henriques 2010).[6]

---

[6] Conrad advises for two algebraic closures of a field '*always* keep track of the choice of isomorphism. In particular, always speak of *an* algebraic closure rather than *the* algebraic closure' (Conrad *n.d.*).

The case of 'the cyclic group of order 6' is also relevant here. We will see later how types of structured types are defined in HoTT. Let us take it, then, that we have a type of groups, *Group*. Now we are to pick out a subtype of this, *CyclicGroupOrder6*, to which end two ways of using dependent sum present themselves. One of these ways is to say that we have a group equipped with a specified single generator of order 6. The second is to say that we have a group and a guarantee of the existence of some (unspecified) single generator of order 6. For either of these two types, all of its elements are isomorphic; however, in the second case, an element, that is, a cyclic group of order 6 but with no specified generator, has a nontrivial automorphism, as we can see from the map sending 1 to 5 in the type formed by $\{0, 1, 2, 3, 4, 5\}$ under addition modulo 6. One should therefore show the same wariness about the employment of 'the' in that case as with 'the algebraic closure of field $F$'. However, if a generator is specified as a part of the structure, then all is well.

Similarly, we could form *Group6*, the type of groups of order 6. This type is not contractible, since there are two connected components corresponding to the two non-isomorphic groups, so we are not allowed to form 'the group of order 6'. This should indicate to us again that we are not dealing with a *generic* form of 'the', as in 'the platypus'. There are 'lawlike' things we could correctly assert about all groups with six elements, and yet we don't say, for example, 'the group with six elements has an element of order 3'. We don't because there are *two* groups of order 6. So although there appears to be a close similarity between the following two sentences:

- There are two bears in Alaska: the grizzly bear and the black bear
- There are two groups of order 6: the cyclic group and the symmetric group on 3 objects

the fact that we can speak of 'the bear' as in 'the bear is an omnivorous mammal', but not 'the group with six elements' is a strong indication that we are employing a *definite* 'the' in the case of groups.

Looking more closely at the *generic* form of 'the', as in:

- The wombat is nocturnal,

we find debate here as to whether the term 'the wombat' is referring to the type or kind of wombats, or else to a typical representative of the type. Clearer evidence for the former interpretation occurs by contrasting pairs, such as

- Edison smashed the light bulb.
- Edison invented the light bulb.

Here, we might say in the second sentence that 'the light bulb' is short for 'the type which is the type of light bulbs'. Edison has designed what it is to be a light bulb. This type, rather than any element, is something that can be patented. It is an instance of that type that can be smashed.

We might treat 'the wombat' similarly, as though it had a characterizable blueprint, one which dictated its nocturnal behaviour. Like Edison inventing a type of artificial light source, we could imagine a divine being creating a species: *God created the wombat*. On the other hand, it is easy to see why people consider there to be a generic individual being referred to. Whereas a type can be invented, it seems odd to describe a *type* as nocturnal rather than an *individual animal*, a member of that type. Were we to look to denote the *typical representative* interpretation of 'the wombat', we could turn to bracket types. As mentioned earlier, for any type, $A$, there is a bracket type $||A||$, which may be thought of as the result of identifying all elements of $A$, if any are present. Some (decidable) properties applicable to wombats won't hold for all wombats, or in other words a map $f : Wombat \rightarrow 2$ need not factor through $||Wombat||$. However, properties that do apply generically to all wombats do so factor through it. $||Wombat||$ is a singleton, as Wombat is an extant species, and any specific wombat, $wombat_1$, provides its element, $|wombat_1|$. Then perhaps when we say 'The wombat is noctural', 'the wombat' should be formed by 'the introduction',

$$the(||Wombat||, |wombat_1|, p),$$

for $p$ a proof of contractibility. Now with the induced arrow to $2$, we have an element of the proposition *The wombat is nocturnal*.

What of contractible, and so true, propositions, taking these in the HoTT proof-irrelevant sense? Well here, while we do not prefix 'the' to a proposition such as 'it is raining', we certainly do say 'the fact that it is raining'. If we wish to retain our 'the introduction' rule here, we might write the type as 'Fact that such and such'.[7] Then we would have 'Fact that it is raining' as a type, and, if it is inhabited, we would designate its element by the term 'the fact that it is raining', which appears to convey well that there is no multiplicity involved, for instance, of warrant for the assertion of the proposition. If 'fact' is considered to be employable only in the case of true propositions, then 'state of affairs' might provide an alternative.

Before moving on, let us clarify what may present itself as a problem under a naïve reading of identity statements relating two definite descriptions, as in the famous case of 'The evening star is the morning star'. While we cannot identify terms belonging to different types, it is reasonable to consider these types as produced by dependent sum

$$\sum_{(x:Star)} Shine\ brightly\ in\ morning(x),$$

and the corresponding one for *evening(x)*. Then while 'the morning star' corresponds to a star (or at least a celestial body), a warrant that it shines brightly in the morning, along with

---

[7] Choosing a case from natural language as here, we encounter the considerations which gave rise to Vendler's advice in his 'Causal Relations' (1967b) and elsewhere to distinguish between facts and propositions. HoTT having been devised by logicians and mathematicians, the mathematical use of 'proposition' as theorem has been employed. Vendler says that one can know a fact, but believe a proposition. As discussed in Chap. 2, it seems that there is dependence not only on $P : Prop$ but also on $p : P$. Perhaps this explains lexical differences: *He knows whether it will rain; *He believes whether it will rain*. Does this explain why we say, '*He knows the neighbours*'? You know a term, not a type.

a guarantee of its uniqueness in this respect, a projection on the first component lands us in the type *Star*. The evening star is treated similarly, and now the two celestial bodies can be compared under the identity criteria for stars (as understood at the time). To put this in natural language, we could say that the star which shines brightly in the morning is the same star as the one that shines brightly in the evening.

In such cases of types being defined by *relative* clauses, we could then propose a second rule called '*relative* the introduction':

$$x : A, B(x) : Type, ((a,b),p) : isContr(\sum_{(x:A)} B(x)) \vdash$$

$$the\ A\ which\ is\ B((a,b),p) \equiv a : A.$$

This makes better sense of how when we say, for instance, "The cat sitting in the basket is Siamese", we mean the property 'Siamese' to apply to elements in the type *Cat*, rather than to those in *Cat sitting in the basket* as a dependent sum. It is worth noting here that some theorists prefer an '*extrinsic*' form of typing rather than our '*intrinsic*' form, often associated with the names *Curry* and *Church* respectively. Extrinsic typing allows for elements to belong to multiple types, so that a particular apple might belong to types *Red Apple*, *Tasty Apple*, *Large Apple*, and so on. For the intrinsic type theorist, on the other hand, an element of *Red Apple* must combine judgement of an apple along with its redness. In such a case, the separation of the whole perception into its components seems plausible, but the extent of similar separation is not clear. Returning to the case in §2.6, when I perceive a man, do I see an animate being which is a man, or merely a man?

I will close out this subsection by observing that 'the' can also be used with plural nouns, as in

- Pat knows the names and the dates of the reigns of the Tudor monarchs.

Here, given a type, *A*, 'the *As*' designates all the elements of *A*. We expect that Pat has the information to hand for all six kings and queens (including Lady Jane Grey). But we also apply such a plural '*the*' less exhaustively in a case such as

- The Romans invaded Britain in 55 BC.

Here we mean representatives of a political collective invaded a foreign territory. If it was an illegal invasion, then we might hold all citizens responsible, whether they themselves crossed the English Channel or not. Later, in Rome, the invasion was celebrated, as it achieved the greater glory of the Empire.

### 3.3.2 Definite Description for Dependent Types

As we have seen, as a development of Martin-Löf type theory, HoTT makes great play of dependent types. It is worth considering, then, how 'the' might be introduced for such types. Let us begin with the most familiar case, where the type depended upon is a set:

$$A : Set, x : A \vdash B(x) : Type.$$

In this case, we can form $x : A \vdash isContr(B(x)) : Prop$, and it may be that we can construct $x : A \vdash (b(x), p(x)) : isContr(B(x))$, which establishes that for any element of $A$, the dependent type is contractible. Then we can form a dependent 'the',

$$x : A \vdash the(B(x), b(x), p(x)) \equiv b(x) : B(x),$$

expressed as 'the $B(x)$'. For example, we might have a set of apartments numbered by a set of numbers, $N$. Then a type depending on $N$ could be $Resident(n)$ for $n$ in $N$. We may subsequently learn that each apartment is single-occupied, so that $Resident(n)$ is a singleton set for every $n$. Then, along with $the(Resident(a))$, or in more familiar terms 'the resident of $a$', for a specific $a$ in $N$, we also have $the(Resident(n)) : Resident(n)$ where $n$ is a variable. We can also use so-called $\lambda$-abstraction from the $b(x)$ above, that is, take it as an expression of (*the resident*)$(n)$, where *the resident* $: \prod_{(n:N)} Resident(n)$ is a function of $N$ picking out the unique resident at each abode.

This makes sense of the use of 'the' for a function picking out an element from non-singleton sets, such as 'the captain of team $x$'. We may think of this as $x :$ *Team* $\vdash$ *the captain*$(x) : Player(x)$. Then $\lambda$-abstraction again produces *the captain* $:$ $\prod_{(x:Team)} Player(x)$.

As in §3.2.1, the above makes sense for general types, as in the case of the product of two types,

$$X, Y : Type \vdash the\,product(X, Y) : Product(X, Y).$$

Again, we may consider *the product* as an element of a dependent product, $\prod_{(X,Y:Type)}$ $Product(X, Y)$.

Further interesting cases arise by allowing dependency on types which are not sets, but rather groups, where standard type-theoretic constructions give rise to types which are groupoids. A centrally important feature of groupoids is their ability to capture 'irregular' quotients. This can be illustrated by a simple example which I first learned from John Baez (see Baez and Dolan 2001). Consider a row of six objects spaced evenly along a line, as shown in Figure 3.1 below. Inverting the line about the midpoint identifies the objects, yielding three pairs.

There is nothing surprising here. We can take ourselves to be demonstrating the simple mathematical fact that $6/2 = 3$. Whether we take this to be three bound pairs or three

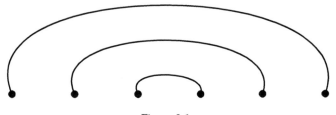

Figure 3.1

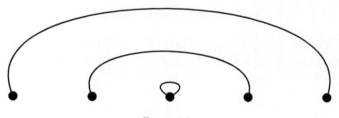

**Figure 3.2**

singletons matters little. Now let us do the same thing with five objects, as in Figure 3.2 below. In this case the central object is unlike the others, in that it will be paired with itself.

The number of classes formed by the identification is three: two pairs and that odd singleton. Here it seems we could be construed as making the evidently false claim that $5/2 = 3$. To repair this, it would appear that we should have that special central object count only for one half. Groupoids give us a way to do precisely this.

For any group, in the mathematical sense, acting on a set, we speak of the 'action' of the group, and represent it by the 'action groupoid'. Think of the elements of the set as points or objects, and arrows going from one object to another are labelled by group elements which send the first object to the second. As mentioned earlier, the cardinality of a groupoid is found by summing for each 'component', or connected cluster of objects, the reciprocal of the size of its 'isotopy group'; essentially, the number of arrows looping from an object to itself. In the case of five objects above, for the object identified to itself in the middle of the row, there are two such arrows. In this case, we have to add $1 + 1 + \frac{1}{2} = \frac{5}{2}$, as desired. Now it is a general result that whatever the group, and whatever its action on a finite set, we will have an arithmetic identity in that the cardinality of the action groupoid is the size of the finite set divided by the order of the group.

For the purposes of physics, again we need more geometric structure, but the basic idea of an action by a group of transformations specifying invariance under symmetries is at work. From 'orbifolds', which might result from a Lie group acting unevenly on a manifold, to the objects of field values for gauge field theory, which need to keep track of gauge equivalences, we find we are treating geometric forms of groupoid. If in the latter case instead we take the simple quotient, the plain set of equivalences classes, which amounts to taking gauge symmetries to be redundant, the physics goes wrong in some sense, in that we can't retain a *local* quantum field theory. Furthermore, in the case of gauge equivalence we don't just have one level of arrows between points or objects, but arrows between arrows, and so on, representing equivalences between gauge equivalences. And this process continues indefinitely to 'higher' groupoids, needed to capture the higher symmetries of string theory (Schreiber 2013). In conceptual terms, the very idea of sameness or identity is being given a much richer interpretation here (Corfield 2003, Chap. 10). This is embodied in the form of the identity types we saw in sections 3.1.1 and 3.1.2.

The action groupoid is in fact a manifestation of something we have already seen. Indeed, in homotopy type theory, the way to express that a set, $A$, is acted on by a group, $G$, is to write it as a dependent type, $* : \mathbf{B}G \vdash A(*) : Type$. The type $\mathbf{B}G$ is a version of the group $G$, but where we're taking it to be a single object with looping arrows labelled by each group element. So $\mathbf{B}G$ has one element, $*$, and $Id_{\mathbf{B}G}(*, *) \cong G$. Then a dependent type, $V$,

$$* : \mathbf{B}G \vdash V(*) : Type,$$

is a type which is equipped with an action by $G$. The single element of $\mathbf{B}G$ is sent to a type, $V$, in $Type$, and the elements of $Id_{\mathbf{B}G}(*, *)$ are sent to automorphisms of $V$, respecting the composition of elements of $G$.

Group actions may be used to characterize simple situations in which there is inherent ambiguity. Imagine that we are in communication with one another and are looking sideways on into a space in which there are three identical balls in a row, but that we are unable to signal to each other which end of the row is which. We can describe ourselves as working in the context $\mathbf{B}S_2$, where $S_2$ is the two-element group. This group acts simply on the row, with the non-identity element reversing the order. Now I could look to use the expression 'the ball on the left' to pick out one ball, but of course you cannot know which of the two ends I mean. On the other hand, I can say 'the ball in the middle' and successfully convey an intended ball. Left and right are not invariant under the group action, whereas being in the middle is invariant in this way. *Middle Ball* is a type in this context and indeed is contractible, allowing me to form 'the middle ball'.

Another way to describe this situation is via the dependent product construction, $\prod_{(*:\mathbf{B}S_2)} Ball(*)$. Recall that elements of a dependent product are functions from elements of the type depended upon to the dependent type. In the case of group actions, there is a single element, $*$, in $\mathbf{B}G$, so a function in the dependent product picks out an element of $V$, but one which has all of the group elements of $G$, leaving it invariant. In our case here, the only invariant element is the middle ball. On the other hand, we might want to speak of being positioned in *the* middle or at *the* end. We can do this by forming the dependent sum, $\sum_{(*:\mathbf{B}S_2)} Ball(*)$. This collects the orbits of that action, in other words a type formed of the elements of $V$, but where whenever a group element, $g$, acts on $v$ to give $v'$, there is an element equating $v$ and $v'$. Imagine here the three balls in a row and, ignoring the trivial arrows, an arrow in each direction between the end balls and a looping arrow at the middle ball. This type has cardinality $1 + \frac{1}{2} = \frac{3}{2}$, in agreement again with the general result that a group, $G$, acting on a set, $X$, generates a dependent sum of cardinality $|X|/|G|$. A little more work is necessary to render it equivalent to a set with two elements, 'middle' and 'end', namely, truncation to the set of its connected components.

Similarly, playing noughts and crosses (tic-tac-toe), I may say 'I like to start in a corner', but it would be reasonable also to say 'I like to start in *the* corner'. Again, the best way to view that latter expression is as a term in (the truncation of) the dependent sum, since my initial play is invariant under the symmetries of the grid.[8] Of course, once I've broken the symmetry by playing in a corner, you can't just say that you would respond to my play in the corner by a play in 'the side square', since there's a difference between an adjacent and a non-adjacent such square.

Working in such *equivariant* contexts, that is, in the presence of a group of symmetries, even though we may not be able to convey unambiguously the identity of elements of a type to one another, we may still agree that the type is a (dependent) set, in the HoTT sense, and

---

[8] The dihedral group of order 8.

as such has a set cardinality. So I can say 'there are three balls' in my original situation above and you will agree. The similar case of two identical balls in an otherwise empty space has generated a considerable debate in metaphysics around the issue of how two things can be non-identical if there are no properties to tell them apart (Black 1952). Some have argued that despite there being no way to pick one out through communicable properties, one ball does differ from the other in the sense that it is identical to that ball itself, while the other ball isn't. If I am looking on, I can distinguish 'this ball' from 'that ball', and think 'this ball is this ball and that ball is not'. However, the other party in this debate sees such a property as illegitimate for purposes of identification.

We now see this philosophical disagreement being illuminated by the type-theoretic understanding of there being a set of cardinality 2 in the context $\mathbf{BS_2}$. I can hear my interlocutor saying 'there's one ball and there's another', and agree with this claim. Their mention of 'one ball' is a term in the context of the symmetries, $* : \mathbf{BS_2} \vdash b(*) : Ball(*)$; it is not a 'one' in the absolute or empty context, where nothing appears to the left of '$\vdash$'. When they then mention a 'different one', I can understand them too, this difference being invariant under the symmetry group. On the other hand, in the empty context where I have bound the free variable, the only types available are the dependent sum $\sum_{(*:\mathbf{BS_2})} Ball(*)$, a set of cardinality 1 and the dependent product $\prod_{(*:\mathbf{BS_2})} Ball(*)$, which is empty. To sum up, there are ways in this framework to say: there are two balls present, there is one kind of thing, but nothing is distinguishable.[9]

## 3.4 The Structure of $A$

When the historian of science, Peter Galison, wrote 'I am uncomfortable with the definite article that begins the title of Kuhn's great work, *The Structure of Scientific Revolutions*' (Galison 1997, p. 60), he evidently meant to say that scientific revolutions had structures of different kinds, or, in other words, that the set of such structures contained more than one element. When mathematicians speak of 'the structure of the natural numbers' or 'the structure of the monster group', they appear to be referring to one such thing.

Philosophers of mathematics have long discussed what is meant by the expression 'the structure of $A$' for a given mathematical entity, $A$. Famously, it is possible to give different constructions within set theory of sets which may be taken to represent the natural numbers (Benacerraf 1965). The structure common to these constructions is then understood by many structuralists to be what the natural numbers are, individual numbers being places in the structure. It is also thought by some of these structuralists that to isolate the structure of any construction there needs to be a way to abstract it from whatever it is that 'carries'

[9] Along similar lines, but in a more intricate setting, HoTT provides an excellent way to understand general covariance in physics (nLab). It is striking how such constructions are written into the very machinery of the type theory. Metaphysicians have also looked to treat indiscernible quantum particles: for example, Lowe (1989). To do this matter full justice along the lines of the present discussion, one would need to consider linear types, corresponding to group representations. Then a 'linear' (in the sense of linear logic) version of HoTT (Schreiber 2014a) should be the right framework to extend the treatment I am giving here.

it, and conditions should be given for when two such abstracted structures are the same (Shapiro 1997, Resnik 1997).

Now HoTT, it is claimed (Awodey 2014), via its univalence axiom, captures what is essentially right about the structuralist position. By use of an 'abstraction principle', Awodey defines a notion of structure through the isomorphism of types:

$$str(A) = str(B) \Leftrightarrow A \cong B. \tag{DS}$$

This definition is merely suggestive. Awodey is speaking informally here, which may give rise to possible misunderstandings. A more fundamental concept than *isomorphism* in HoTT is *equivalence*, which, as we saw in §3.1.3, can be formulated in the language of HoTT in terms of maps with suitable properties (UFP 2014, §2.4, Chap. 4).[10] Equivalence provides the right identity criterion generally, as the univalence axiom pronounces. Also, the biconditional ($\Leftrightarrow$) is not part of the syntax of HoTT. It could perhaps appear in HoTT as a notational variant of equivalence in the case of two mere propositions, and yet $A \cong B$ as a type is a variant of the type of equivalences, $A \simeq B$, and in general not a mere proposition.

Awodey concludes, after a discussion of the univalence axiom (UA), which says of two types that if they are equivalent as defined in HoTT, then they may be considered equal as elements of the type of small types:

> observe that, as an informal consequence of (UA), together with the very definition of 'structure' (DS), we have that two mathematical objects are identical if and only if they have the same structure:
>
> $$str(A) = str(B) \Leftrightarrow A = B.$$
>
> In other words, mathematical objects simply *are* structures. Could there be a stronger formulation of structuralism? (Awodey 2014, p. 12)

In other words, taking HoTT as our foundation, all constructions are already fully structural.

This conclusion seems to me to be correct, but here I shall adopt a different argument strategy by examining whether HoTT itself can tell us a little more about such locutions as 'the structure of $A$', '$A$ and $B$ share the same structure' and 'places in the structure'. Rather than invoking the Fregean notion of an abstraction principle, as Awodey does, I shall propose what appear to be the only plausible definitions within HoTT itself of the relevant terms.

I shall not engage here in a close reading of the wide array of existing structuralist positions. The main point of this section is to show that, working in HoTT, the kinds of concern that date back to Benacerraf largely dissolve. With our new-found ability to express uniqueness up to canonical equivalence by definite description, the motivation to seek some single entity commonly related to two structurally equivalent entities is removed.

---

[10] 'In general, we will only use the word *isomorphism* (and similar words such as *bijection*, and the associated notation $A \cong B$) in the special case when the types $A$ and $B$ "behave like sets" ' (UFP 2014, p. 78).

Our reconstruction of 'the structure of $A$' essentially requires our generalized 'the', as applying not only to sets but to any types.

Two plausible options present themselves according to different naming conventions for types that we saw in §3.2. Recall that we have (1) *Circle* as the type which behaves like the circle, (2) *Natural number* or $\mathbb{N}$, the type of natural numbers. An element of (1) is not a circle, where an element of (2) *is* a natural number. In the case of 'the structure of $A$', then, we may mean:

1. The type which behaves like the structure of $A$.
2. The unique element up to equivalence of a type *Structure of A*.

Option (1) needs further unpacking. Perhaps we might recast it as 'the type which behaves structurally like $A$'. But then this seems to be no different from '$A$' itself. If so, 'the structure of' is the identity map on types, and completely redundant. This does tally with Awodey's solution, where since $str(A) :\equiv A$, we would have $str(A) = str(B)$ as definitionally equal to $A = B$, and so, trivially, equivalent:

$$str(A) = str(B) \simeq A = B.$$

Then we never need utter 'the structure of' again.

Let's now pursue option (2). To be in a position to define 'the structure of', we will first consider the expression 'structure of' as applied to a type in the system. Together with our analysis of definite description in §3.2, we will then be able to interpret 'the structure of $A$'. Finally, we consider 'places in' such a structure, and extend these analyses to structured types.

So, if we agree with the analysis of §3.2.1, then to be able to say 'the structure of $A$' by 'the introduction' we must already have (a) formed a type, *Structure of A*, and (b) established that it is contractible. Now, one plausible candidate for 'Structure of $A$' is the type

$$Structure(A) :\equiv \sum_{(X:\,\mathcal{U})} Equiv(A, X),$$

where, as usual, $\mathcal{U}$ is the type universe of small types. This is an eminently reasonable choice, since elements of this type are types equipped with an equivalence with $A$; we might say 'types-structured-as-$A$'. What is required now is to establish the contractibility of this type of such types. Intuitively, this should be clear as contraction can take place to $A$, as it were, along the given equivalences. But, of course, a proof in HoTT requires use of its technical apparatus, which I will briefly sketch.

Straight off, we have an element of that type to hand, namely $(A, Id_A)$. $A$ is structured as $A$, as witnessed by its identity map. Then to establish $isContr(Structure(A))$ we also need for every $B : \mathcal{U}$ and $f : Equiv(A, B)$ a canonical way to identify $(A, Id_A)$ and $(B, f)$. What such an identity amounts to in the case of a dependent sum is a path in the base type; here, that is one in $\mathcal{U}$ between $A$ and $B$, and a path over this one in the total space of equivalences to $A$. For the former we use the path that the equivalence furnished by the univalence axiom makes correspond to $f$. The effect of transporting $Id_A : Equiv(A, A)$ in the total space will then be $f \circ (Id_A) = f : Equiv(A, B)$. Let us call this process of identification $p$.[11]

Without any obvious non-equivalent alternatives for 'Structure of $A$', let us pursue this choice by forming the term 'the structure of $A$'. In its full glory, it is

$$the(Structure(A), (A, id_A), p) \equiv (A, id_A) : Structure(A).$$

Dropping $p$, we find that 'the structure of $A$' is $(A, id_A)$. Notice that the component $id_A$ is playing a role here. We should note that, having constructed the type $A$, were we to construct an element, $g : Equiv(A, A)$, which is not equal to $id_A$, then we could equally use $(A, g)$ to witness the contractibility of $Structure(A)$. This would require a modification to $p$, but to the extent that this component is not mentioned, we might equally well say that $(A, g)$ is 'the structure of $A$', or indeed any $(B, f) : Structure(A)$. An element of the type is an entity, $B$, structured as $A$, as witnessed by an equivalence, $f$. Any such element has trivial identity type with any other, $Id_{Structure(A)}((B_1, f_1), (B_2, f_2))$. This is very much like the case we described earlier of the product of types, where we needed not just a type but also extra information, such as its projections. If I do not include the extra information, here the way, $f$, that some type, $B$, is equivalent to $A$, then there need not be only one way that $B$ shows itself to be structured as $A$.

Now what does it mean to say that $A$ and $B$ have the same structure? Well, one might expect that it means to indicate an identity between two elements, 'the structure of $A$' and 'the structure of $B$'. As in the case of the morning star and the evening star, naïvely read they are elements of different types and so not to be directly compared, but like that example we can project to the first component, that is the type that the dependent types are depending upon; here, the universe $\mathcal{U}$. Then the identity of the elements amounts to an identity between types $A$ and $B$ in $\mathcal{U}$, or in other words equivalence between the types.

Alternatively, we might define a $\mathcal{U}$-dependent type '$X$ has the same structure as $A$' $\equiv Equiv(A, X)$. Then consider by $\lambda$-abstraction the term $\lambda X.Equiv(A, X)$, which in words we might say designates 'has the same structure as $A$'. Now, 'has the same structure as $B$' is an element of the same type, and we can ask for their identity type. This can easily be shown to be equivalent to $Equiv(A, B)$.

I mentioned another approach to definite description as the result of applying a function, as in 'the captain of team $t$'. Here we might think there is a function from the type of types, $\mathcal{U}$, to some type of structures, $Structure$, fibred above it. The evident choice for $Structure$ is

$$\sum_{(X:\mathcal{U})} Structure(X) \equiv \sum_{(X,Y:\mathcal{U})} Equiv(X, Y),$$

in which case we have a similar solution to the one above in that 'the structure of' is found to be a function in $\prod_{(X:\mathcal{U})} Structure(X)$, which sends $A$ to $\langle A, (A, id_A) \rangle$.[12]

---

[11] We could also work with an equivalent type: $Structure(A) \equiv \sum_{X:\mathcal{U}}(A = X)$. Then Lemma 3.11.8 of UFP 2014 gives us contractibility.

[12] Since the fibres are contractible, $Structure$ is equivalent to $\mathcal{U}$ (UFP 2014, Lemma 3.11.9).

### 3.4.1 Places in a Structure

Some structuralist philosophers of mathematics have referred to 'places' or 'positions' in a structure (Resnik 1997, Shapiro 1997): for instance, to refer to particular natural numbers in the structure, that is, the natural numbers. This is to indicate elements in what results from a process which abstracts away from different presentations of the 'same structure'. Let us see what it is possible to express within HoTT.

Well, 'places in the structure of $A$' suggests that we form a type which depends on $Structure(A)$. There doesn't appear to be much choice here other than

$$(X,f) : Structure(A) \vdash PlacesIn(X,f) \equiv X : Type.$$

It would be very natural, then, to form the dependent product to allow the collection of coherent choices of element of $A$ along with their corresponding elements in each type structured as $A$, according to the specified equivalence:

$$\prod_{((X,f)\,:\,Structure(A))} PlacesIn(X,f).$$

We might pronounce this 'Places in $A$-structured types'. This type can easily be shown to be equivalent to $A$, since a choice, $a : A$, determines an element, $f(a) : B$, for each type-structured-as-$A$, $(B,f)$, and a choice of place in $A$-structured types delivers an element of $A$ when applied to $(A, id_A)$.

### 3.4.2 Types Equipped with Structure

Of course, we don't just talk about plain types, but also about monoids, groups, vector spaces, and so on. Consider one of the simplest cases, the semigroup structure. This merely requires that there be an associative binary multiplication on the type. Following definition 2.14.1 of UFP (2014),

$$SemigroupStr(A) :\equiv \sum_{(m:A\to A\to A)} \prod_{(x,y,z\,:\,A)} m(x,m(y,z)) = m(m(x,y),z).$$

Now, a semigroup is a type together with such a structure:

$$Semigroup :\equiv \sum_{(A\,:\,\mathcal{U})} SemigroupStr(A).$$

Then for a particular $(A,m,a) : Semigroup$, where $a$ is a proof of the associativity of $m$, we can define

$$Str(A,m,a) :\equiv \sum_{((X,y,z)\,:\,Semigroup)} f : Equiv_{Semigroup}((A,m,a),(X,y,z)),$$

where $Equiv_{Semigroup}$ requires of an element that it is an equivalence between underlying types and that it transports the semigroup structure correctly. Once again, this results in a contractible type as witnessed by $(A, m, a, id_{(A,m,a)})$, which element we may then call 'the structure of the semigroup $A$'. Places in $(A, m, a)$-structured semigroups will again amount to $A$.

### 3.4.3 The Complex Numbers

We can put together the constructions treated above to handle the case of the complex numbers. The issue at stake here is whether, contrary to what makes mathematical sense, some forms of structuralism are forced to identify the two square roots of $-1$ in the field of complex numbers, $i$ and $-i$, given that the nontrivial field automorphism, conjugation, maps them to each other, that we cannot distinguish them in terms of the real-valued polynomials they satisfy, and so on.[13]

How to introduce $\mathbb{C}$ in type theory? Of course, there will be many ways to do so, but we can distinguish two styles of definition, as 'particular' and 'abstract' types.[14] For example, we can form a particular type of complex numbers from the bottom up as ordered pairs of reals, with specified addition and multiplication, and so on. These reals in turn will have a particular structure depending on how they have been defined (see Chap. 11 of UFP 2014). On the other hand, as an abstract type, we can construct a type to which $\mathbb{C}$ as a whole belongs: for instance, the type, $\mathcal{A}$, of algebraically closed fields of characteristic zero and cardinality of the continuum. A concrete construction of a particular type, $\mathbb{C}$, then becomes a proof that $\mathcal{A}$ is inhabited.

These specifications carry different information, the difference being very much like that between the two ways of specifying the cyclic group of order 6 in §3.2.3. In the case of the particular type, any $z : \mathbb{C}$ may be decomposed into its real and imaginary parts. Now $\langle 0, 1 \rangle$ and $\langle 0, -1 \rangle$ are two different elements, both of which square to $-1$. On the other hand, in the case where we assume $\mathbb{C} : \mathcal{A}$, we don't have the means to individuate the two square roots, and yet the subtype of elements squaring to $-1$ is of cardinality 2. There are two such places in the structure. $\mathcal{A}$ is equivalent to $\mathbf{B}Aut(\mathbb{C}) \equiv \mathbf{B}S_2$ (recall this notation from §3.2.4) with a nontrivial structured auto-equivalence which exchanges the two square roots. We might consider the strict condition for the use of 'the' in 'the complex numbers' to require us to break the symmetry by specifying one root as $i$, although in practice mathematicians will often say '$\mathbb{C}$ is the algebraically closed field of characteristic zero and cardinality the continuum'. In any case, the situation is very much like the cyclic group of order 6 or the two identical balls, and presents no difficulty to a type-theoretic viewpoint.

## 3.5 Conclusion

One conclusion to draw from this chapter is that from the perspective of HoTT, little is gained by explicit use of the word 'structure' in the sense of 'the structure of $A$'. Types and

---

[13] See, for example, Keränen (2001, 2006) and other contributions to MacBride (2006), and Nodelman and Zalta (2014).

[14] I am indebted to Mike Shulman's discussion 'From Set Theory to Type Theory' (2013).

structured types in HoTT just *are* structures that do not need to be abstracted from an underlying set-like entity. HoTT simply is a 'synthetic theory of structures' (Shulman 2017). The proper treatment of structure comes along for free and need not be explicitly mentioned. In this sense, then, HoTT should be viewed very favourably by stucturalists. In Brandomian terms, we could take HoTT to be *making explicit* the equivalence-invariant reasoning of mathematicians.

If establishing that talk of the structure of a type is redundant counts as a result which closes off a certain line of enquiry for the structuralist, on the way to it we have seen something more positive.

1. The analysis of the word '*the*' in terms of its introduction rule shows that HoTT has something to teach us about the classic philosophical topic of definite descriptions. We have seen that it provides a rationale for mathematicians' use of a generalized 'the' in situations where it appears that they might be referring to more than one entity.

2. More generally, we were able to make useful sense of several issues concerning type and identity. HoTT promises to be an important tool for philosophers of language and metaphysicians.

3. Our analysis of 'the' employed a principle that may prove of lasting importance:

   (*Treat all types evenly*) Any time we have a construction which traditionally has been taken to apply only to sets or only to propositions, then since in HoTT these form just a certain kind of type, we should look to see whether the construction makes sense for all types.

   Further instances of this principle are not hard to find. If we generally take modal operators, such as 'it is necessarily the case that . . . ', to apply only to propositions, we should look to see whether there is anything to prevent a more general construction applying to all types. We will look to do just this in Chap. 4.

It is surely intriguing that a newly proposed foundational language for mathematics displays the potential to speak to issues within philosophy in general, and is not confined to the domain of philosophy of mathematics.

# 4

# Modal Types

We now proceed to take the final step in our journey towards modal homotopy type theory. Analytic philosophers have put modal *logic* to extensive use in their exploration of so-called alethic, epistemic, doxastic, deontic, temporal and other modalities. These modalities typically qualify ways in which a proposition may be said to be true, as with

- It is necessarily the case that ...
- It is known to be the case that ...
- It is obligatory that ...
- It will be the case that ...

The relevant logical calculus is shaped by, and in turn shapes, reflection on what is imagined to be the philosophical content of these concepts. So philosophers might consider the differences, if any, between *physical, metaphysical* and *logical* necessity and possibility. Such discussions will often involve consideration of what has played the role of semantics for these logics; in particular, possible world semantics. For instance, '*P* is necessarily the case' might be taken to mean '*P* holds in every metaphysically possible world'.

Saul Kripke had given a warning concerning his introduction of possible worlds:

> The apparatus of possible worlds has (I hope) been very useful as far as the set-theoretic model-theory of quantified modal logic is concerned, but has encouraged philosophical pseudoproblems and misleading pictures. (Kripke 1980, p. 48)

All the same, many have decided to overlook his concerns, encouraged by the example of David Lewis (1986). While not every metaphysician has adopted such a concretely realist attitude towards these worlds as Lewis, numerous criticisms even of their instrumental usage have been launched from a range of positions, from the Wittgensteinian to the constructive empiricist. For Hacker (2010), Wittgenstein rightly understood necessary truths as merely the expressions of the normativity of rules—rules of grammar, mathematics or whatever: 'Necessary propositions exhibit neither factual or super-factual

*Modal Homotopy Type Theory: The Prospect of a New Logic for Philosophy.* David Corfield, Oxford University Press (2020). © David Corfield. DOI: 10.1093/oso/9780198853404.001.0001

("meta-physical") nor ideational (psychological) truths, but rather conceptual connections' (p. 20). Hacker maintains that there are no possible worlds beyond our actual world. For van Fraassen (2002), philosophers toying with these possible worlds are engaged in a palid imitation of science, where speculative claims are made without any form of check beyond the supposed coherence of the account offered: 'The world we live in is a precious thing; the world of the philosophers is well lost for love of it' (p. 18).[1]

But even if one has sympathy for complaints of this kind, it is worth observing that computer scientists have latched onto modal logic and run with it, including a good number happy to work on developments of Kripke's semantics. They do this *both* by taking up the modalities of the philosophers and putting them to different uses—for instance, temporal logic in model-checking and epistemic logic in multi-agent systems—*and* also by devising new modalities of their own. So, in regard to the latter, modalities have been defined to represent security levels and computational resources, and more generally, what they term *effects* and *coeffects*, relating to features of the context in which programs are executed beyond mere input–output pairings.

Computer scientists are also technically inventive. Where philosophers are still largely working exclusively with propositional modal logics (K, S4, S5, and so on), first-order extensions and Kripke models for semantics, computer scientists also employ substructural logics, coalgebra, labelled transition systems, descriptive frames and bisimulations, where these topics are often given a category-theoretic treatment.

With fewer tools at their disposal, the task of philosophers looking to engineer a useful form of *first-order* modal logic is not made easy by the lack of clear-cut conceptual constraints placed upon them. Indeed, Kishida makes just this point in his contribution to *Categories for the Working Philosopher* (Landry 2017):

> Modal logicians have devoted the overwhelming majority of their inquiries to propositional modal logic and achieved a great advancement. In contrast, the subfield of quantified modal logic has been arguably much less successful. Philosophical logicians—most notably Carnap, Kripke, and David Lewis—have proposed semantics for quantified modal logic; but frameworks seem to keep ramifying rather than to converge. This is probably because building a system and semantics of quantified modal logic involves too many choices of technical and conceptual parameters, and perhaps because the field is lacking in a good methodology for tackling these choices in a unifying manner... the essential use of category theory helps this situation, both mathematically and philosophically. (Kishida 2017, p. 192)

He then goes on to provide there a rich account of first-order modal logic. What I shall be working towards in this chapter, on the other hand, is the more ambitious target of a modal version of homotopy type theory, naturally still in a category theory-guided way.

We begin by tracing a path which can help us understand why modalities are often conceived in terms of variation over some collection, often construed by philosophers to

---

[1] Elsewhere he writes, 'However plausibly the story begins, the golden road to philosophy which possible-world ontologies promise, leads nowhere' (van Fraassen 1989, p. 93).

be a set of worlds. One key finding, however, is that it is not this variation as such that matters, but rather the properties possessed by *adjoint* operators arising from such variation. These properties are encountered in a broader range of situations, and are now forming the basis of a powerful new approach to modal type theory, mentioned in the final section of this chapter. I believe that *computational trinitarianism* is pointing us very strongly in this direction.

As for the expected payoff, modal type theories are already being deployed extensively in computer science. Moreover, as we will see in Chap. 5, they can also make good sense of developments in current geometry. Furthermore, *linear* dependent type theories are now being developed, and are expected to provide a syntax for the forms of monoidal category used in quantum physics (Schreiber 2014a).

As for philosophy, for analytic metaphysics, modal HoTT can help us to think through options on modal counterparts profitably. It also offers a range of novel lines of investigation, such as a way to think beyond modal propositions to elements of modal types, such as '*necessary* steps' and '*possible* outcomes'. But in view of the type theory-inferentialism relationship that has been noted at points through this book, we might also expect to make common cause with inferentialist perspectives on modality; in particular, what Robert Brandom denotes as the *Kant–Sellars thesis about modality*:

> in being able to use nonmodal, empirical descriptive vocabulary, one already knows how to do everything one needs to know how to do in order to deploy modal vocabulary, which according can be understood as making explicit structural features that are always already implicit in what one *does* in describing. (Brandom 2015, p. 143)

The uses of nonmodal vocabulary being made explicit by modal language include those where one describes how the state of something would be under certain kinds of variation of its current situation. As we shall now see, such variation plays a key role in forming modalities for type theory.

## 4.1 Modalities as Monads

First let us consider the *alethic* modalities:

- It is necessarily the case that ...
- It is possibly the case that ...

Modal operators such as these are often considered to possess certain structural features irrespective of the nature of the associated proposition. For example, a standard reading of *possibility* admits the following implications:

- $p$ implies possibly $p$.
- Possibly possibly $p$ implies possibly $p$.

When symbolized as entailments, $p \vdash \Diamond p$ and $\Diamond\Diamond p \vdash \Diamond p$, a category theorist will immediately be put in mind of a key construction known as a *monad*. A similar analysis of necessity indicates the dual concept, a *comonad*.

To understand modalities from a type-theoretic perspective we will need to make sense of monads and comonads and of how they arise within the context of category theory, but let us ease our way into this material by looking at a simple case. So consider the situation in which there is a type of dogs and a type of people, and where each dog is owned by precisely one person. Then the mapping

$$owner : Dog \rightarrow Person$$

allows any property of people to be transported to a property of dogs: for instance,

*Being French $\mapsto$ Being owned by a French person.*

We may say that taking properties of *Person* as elements of $2^{Person}$, the 'owner' function induces a mapping

$$owner^* : 2^{Person} \rightarrow 2^{Dog},$$

from being a kind of person to being owned by that kind of person.

Now, we may order these domains of properties as partially ordered sets, where the ordering corresponds to inclusion: for instance, 'pug' is included in 'toy dog' and 'being French' is included in 'being European'. However, we cannot necessarily invert this mapping to send a property of dogs, say, 'being a pug', to a property of people. Indeed, there is no reason that being a certain kind of dog should correspond to being owned by a certain kind of person.

We may try

*Pug $\mapsto$ Owning some pug* ,

but then the composite results in

*Pug $\mapsto$ Owning some pug $\mapsto$ Owned by someone who owns a pug.*

However, people may own more than one breed of dog. A poodle who shares a home with a pug will count as a dog owned by someone who owns a pug.

Since this doesn't work, let's try

*Pug $\mapsto$ Owning only pugs.*

However, this leads to

*Pug $\mapsto$ Owning only pugs $\mapsto$ Owned by someone owning only pugs.*

Yet again, this is not an inverse, not all pugs being owned by single breed owners. The pug sharing its home with a poodle is a case in point.

Despite the failure to find an inverse, my suggestions, to be denoted respectively $\sum_{owner}$ and $\prod_{owner}$, are in some sense the best approximations to one. Too loose an approximation first time; too constrictive an approximation second time. But even though they are only approximations to inverses, we can show that these mappings between dog and people properties may be used to make inferences. Say we have two people, $D$ and $P$, trying to establish a relationship between their fields, where $D$, the dog expert, can only think in terms of dogs and their properties, and $P$, the people expert, can only think in terms of people and their properties. Entailments for one expert may be translated to entailments for the other. For instance, when $P$ establishes that being French implies being European, $D$ can know that being owned by a French person entails being owned by a European. And when $D$ establishes that all pugs are toy dogs, then $P$ knows that an owner of only pugs is an owner of only toy dogs, and similarly for 'only' replaced by 'some'. So their inference patterns are in this sense reflected in the other's.

But we might ask for more. Even though there is no process to translate properties of concern to each other faithfully across the divide, the experts are still able to establish jointly a relationship between a dog property and a person property. Let's say that $D$ chooses the property *pug* and $P$ chooses *being French*. $D$ starts listing names of dogs which are pugs. $P$ doesn't understand anything about these matters, but what appears to him are the results of the *owner* function, that is, a series of people's names, which are in fact the owners of a pug. Now $P$ might check to see if each such owner is French. Then they will jointly assert or deny the following claims.

- $D$: All pugs are owned by a French person (whatever such a thing is).
- $P$: Any person who owns a pug (whatever that is) is French.

A similar analysis works for the other mapping:

- $D$: All dogs owned by a French person (whatever such a thing is) are pugs.
- $P$: A French person owns only (if any dog at all) pugs (whatever they are).

What we have here is an *adjoint triple*, $\sum_{owner} \dashv owner^* \dashv \prod_{owner}$, acting between the partially ordered set of dog properties and the partially ordered set of people properties, treating these as categories. To make contact with the opening comments of this chapter on the properties of modal operators, we now need to look more closely at the result of composing pairs of adjoints. As we have seen, taking a predicate of dogs and applying the left adjoint followed by the *owner*$^*$ mapping yields a self-map of dog properties:

*Pug* $\mapsto$ *Owning some pug* $\mapsto$ *Owned by someone who owns a pug.*

Certainly, in a world where all dogs are owned by some person, if a dog is a pug, then it is owned by someone who owns a pug. However, the opposite condition does not hold, since people may own more than one breed of dog. On the other hand, iteration of the construction is *idempotent* in the sense that 'being owned by someone who owns a dog which is owned by someone who owns a pug' is equivalent to 'being owned by someone who owns a pug'. Structurally the resemblance to possibility is clear. Being owned by someone who owns a pug is being construed as though you 'might have been a pug'. So 'owned by someone who owns' resembles the possibility operator, both being monads. This is the best $D$ can do in having $P$ help him to establish a consequence of possession of a dog property. $P$ has information concerning a weaker property. If you learn via $P$ that a particular dog is co-owned with a pug, there's still a chance it may be a pug.

Similarly, we can form a *dual* version where we begin with the *right* adjoint. In our case, for example, we can tell a comparable story about the comonad on the category of properties of dogs, generated by the right adjoint of *owner**:

$$Pug \mapsto Owning\ only\ pugs \mapsto Owned\ by\ someone\ owning\ only\ pugs.$$

Again, not all pugs are owned by single-breed owners, so we have an implication from 'being owned by someone who owns only pugs' to 'being a pug', but not in the other direction. On the other hand, 'being owned by someone who owns only pugs' is equivalent to 'being owned by the owner of a dog owned by someone who owns only pugs'. Evidently, this operation is acting like necessity. If you and all your co-owned fellow dogs are pugs, then you're 'necessarily' a pug. Now, 'owned by someone who owns only' is seen to resemble necessity as a comonad. Again, $P$ is doing their best to provide information from which $D$ can look to conclude possession of a property. $D$ may learn via $P$ consequences of a specific dog being one of a group of dogs, all of which are pugs. $P$ can provide information about a stronger property.

In sum, these constructions applied to our pug case are:

- $\Diamond_{owner}$ : *Pug* $\mapsto$ *Owning some pug* $\mapsto$ *Owned by someone who owns a pug*.
- $\Box_{owner}$ : *Pug* $\mapsto$ *Owning only pugs* $\mapsto$ *Owned by someone owning only pugs*.

We have equivalents of

- $P \rightarrow \Diamond P$.
- $\Box P \rightarrow P$.

As we saw above, we also have equivalents of

- $\Diamond\Diamond P \rightarrow \Diamond P$.
- $\Box P \rightarrow \Box\Box P$.

Now, as I mentioned in §1.1, partially ordered sets can be considered as categories enriched in truth values. Unsurprisingly, then, adjoint triples are commonly encountered operating between ordinary categories, where they also generate monads and comonads. As we should expect with central category-theoretic constructions, *monads* appear throughout mathematics. Let's take a look at how these arise. For instance, consider a set, $S$, along with the associated set, $M(S)$, of finite strings of elements of $S$. Then there are two natural mappings:

- $i : S \rightarrow M(S)$, which sends an element $s \in S$ to the string of length one, $\langle s \rangle$.
- $m : M(M(S)) \rightarrow M(S)$, which sends a string of strings of elements of $S$ to the concatenated string, for example, $m : \langle\langle pqr\rangle\langle st\rangle\langle uvw\rangle\rangle \mapsto \langle pqrstuvw\rangle$.

Notice that $M$ is behaving very much like $\Diamond$, as is made apparent by representing propositions as objects and entailments as arrows: $p \rightarrow \Diamond p$ and $\Diamond\Diamond p \rightarrow \Diamond p$.

This data forms a *monad*, since we have a category, here the category of sets, *Set*, a functor from the category to itself,[2] here $M$, with a unit, $i_S : S \rightarrow M(S)$, and a multiplication, $m_S : M(M(S)) \rightarrow M(S)$, for each $S$ in *Set*, satisfying a number of equations. In this case of strings of elements of a set, some of these equations concern obvious properties of the concatenation of singleton strings. Other equations tell us that concatenating strings of strings of strings of elements, it does not matter whether we begin concatenating at the inner or the outer layer.

We pay less attention to this extra structure in the logical case since we generally take an arrow between propositions to represent entailment, rather than a specific entailment. So the inference from $\Diamond\Diamond\Diamond P$ to $\Diamond P$ may be derived in two ways, the first step either being to reduce the first two $\Diamond$s to one, or alternatively to reduce the second and third of them to one.

All monads on a category, $\mathcal{C}$, come from a composition of two adjoint functors between $\mathcal{C}$ and some other category, $\mathcal{D}$, the left adjoint followed by the right. In general, this composition occurs in several different ways, in the sense that non-equivalent choices of $\mathcal{D}$ are possible. In the case of our strings of elements, $M$ may be taken as the composite of a pair of adjoint functors between the category of sets and the category of monoids, a monoid being a set equipped with an associative binary operation and a unit for this operation. The left adjoint is the *free* functor which sends a set to the monoid it freely generates, the identity element being the empty string. The right adjoint is the forgetful functor, which sends a monoid to its underlying set.[3]

It seems, then, as though familiar modal operators such as 'possibly', behaving as they do like a monad, should equally arise from an adjunction. Let's now show this by returning to our dog scenario, but where before I used the *owner* map, I now work with a different map. The constructions I detailed above concerning dog ownership work for any map between sets, and so in particular for the terminal map ($Dog \rightarrow \mathbf{1}$). Induced mappings must send dog properties to 'properties of $\mathbf{1}$'. We can make sense of the latter—properties of $\mathbf{1}$ are just propositions. The equivalent of the *owner*\* map is a map which sends a proposition, $Q$, to the property of dogs—any dog if $Q$ is true, no dog if $Q$ is false. The result of this *pull-back* process, which we might denote $Dog^*Q$, is equivalent to attaching a copy of the constant proposition, $Q$, to each dog. Then attempts at inverses to this mapping, left and right adjoints, would

---

[2] Also called an *endofunctor*.
[3] So-called *algebras* for this monad are monoids.

send a property of dogs such as 'being a pug' to the familiar constructions of quantified statements: 'Some dog is a pug' and 'All dogs are pugs'.

Now the step to the possible worlds of modal logic is simple to make.[4] One common philosophical interpretation of *necessarily* and *possibly* is in terms of a collection of *possible worlds* of which our *actual world* is just one element. So let $W$ be the type of all possible worlds. Any specific choice of $W$ may be taken as specifying what is to be understood as a possible world. Under this interpretation, a proposition that depends on $W$ is necessarily true if it is true in all possible worlds, and possibly true if it is true in some possible world. Note, however, that these operations change the dependence from $W$ to non-dependence, or dependence on **1**. In other words, if a proposition, $P(w)$, depends on $W$, so that it may be true in some worlds and false in others, then $\exists_{w:W}P(w)$ and $\forall_{w:W}P(w)$ no longer depend on $W$. But the idea of a necessity and a possibility modality is to send a proposition in some context to a proposition in the *same* context so that they may be compared. We should be able to say, for instance, that $\Box P$ implies $P$, and so on. Thus we need to make $\exists_{w:W}P(w)$ and $\forall_{w:W}P(w)$ into propositions that again depend on $W$, even if they now depend trivially on $W$.

We do this by using the *pull-back* for $W \to \mathbf{1}$, generating what we might call a constant world-dependent proposition. The result of this $W^*$ applied to a non-dependent proposition, $Q$, is to set a copy of $Q$ sitting above every world, so that we have an identity projection on the second component, $Q \times W \to W$. Now, the composite monad and comonad are as follows:

$$(\underset{W}{\Diamond} \dashv \underset{W}{\Box}) := \left( \left( W^* \circ \underset{W}{\exists} \right) \dashv \left( W^* \circ \underset{W}{\forall} \right) \right) \;:\; \mathbf{H}_{/W} \longrightarrow \mathbf{H}_{/W},$$

taking $\mathbf{H}_{/W}$ for the moment as the category of world-dependent *propositions*, with implications as arrows. With this, if $P \in \mathbf{H}_{/W}$ is a proposition about terms $w$ of $W$ (a $W$-dependent type), then

- $\underset{W}{\Diamond}P(w)$ is true at any $w$ precisely if $\underset{W}{\exists}P(w)$ is true, hence if it is the case that $P(w)$ is true for some $w$;

- $\underset{W}{\Box}P(w)$ is true at any $w$ precisely if $\underset{W}{\forall}P(w)$ is true, hence if $P(w)$ holds for all $w$.

It may appear to be the case that the further operation provided by applying $W^*$ is unnecessary, but it is crucial for a proper modal HoTT treatment.

Thus we are given one syntactic formalization of the informal meaning of necessity and possibility. The natural semantics for these base change operations is a generalization of the simple traditional possible-world semantics of propositional necessity and possibility modalities. Notice that we arrived at these constructions without the usual device of using negation. In classical modal logic, the operators are interdefinable as $\neg\Box P = \Diamond\neg P$. Here, however, we defined them independently. Moreover, with this formalization, the modal operator $\Diamond_W$ is left adjoint to $\Box_W$ and hence together they form an *adjoint modality*. Indeed,

---

[4] This formulation was worked out on the nLab.

when an adjoint triple is used to form a pair of modalities in this way, they are always in turn adjoint to one another, expressing their 'opposition':

$$\Diamond P(w) \to Q(w) \Leftrightarrow P(w) \to \Box Q(w).$$

In words, if the possibility of $P$ entails that $Q$ holds at this world, then were $P$ to hold at this world then $Q$ would necessarily be the case. Choosing $Q$ equal to $P$, we see that a proposition sits between the images of the two operators:

- necessarily true, true, possibly true

following the pattern of

- everywhere, here, somewhere.

The modal adjoints additionally furnish us with an equivalent for axiom(B) from standard modal logic:

$$Hom(\Diamond P(w), \Diamond P(w)) \cong Hom(P(w), \Box\Diamond P(w)).$$

Also, we have (5)

$$Hom(\Diamond\Diamond P(w), \Diamond P(w)) \cong Hom(\Diamond P(w), \Box\Diamond P(w)).$$

This is evidently an S5 form of modal type theory.

A general property of adjoints is that they preserve certain kinds of categorical structure. Left adjoints preserve sums (and colimits in general), while right adjoints preserve products (and limits in general). Examples of this preservation are that

- possibly $P$ or $Q \leftrightarrow$ possibly $P$ or possibly $Q$
- necessarily $P$ and $Q \leftrightarrow$ necessarily $P$ and necessarily $Q$

While base change-adjunctions are essentially unique and not free to choose, there is a genuine choice in the above given by the choice of context $W$. This is reflected in the subscripts of $\Diamond_W$ and $\Box_W$ above. It is the choice of this $W$ that gives different kinds of possibility and necessity. More generally, there is in fact not just a choice of a context, but of a morphism of contexts, reflecting what is often called *accessibility* of possible worlds.

This construction resembles the dog-owner case better if we consider an equivalence relation on *Worlds*, represented by a surjection, $\omega : W \to V$, where the condition that two worlds be related, $w_1 \sim w_2$, is given by $\omega(w_1) = \omega(w_2)$. Now, necessarily $P$ holds at a world, $w$, if $P$ holds at all worlds related to $w$, that is, all worlds with the same image, $\omega(w)$, in $V$ as $w$. With our axiomatization via base change, it is simple to adjust the previous construction to this relative case. Indeed, instead of base change along the map to the unit type $W \to \mathbf{1}$, we now consider base change along the surjection, $\omega : W \longrightarrow V$.

$$(\exists_\omega \dashv \omega^* \dashv \forall_\omega) \; : \; \mathbf{H}_{/W} \; \underset{\underset{\exists_w \,:\, \omega^{-1}(-)}{\longrightarrow}}{\overset{\overset{\forall_w \,:\, \omega^{-1}(-)}{\longrightarrow}}{\underset{\omega^*}{\longleftarrow}}} \; \mathbf{H}_{/V}.$$

Then we set

$$\left(\underset{\omega}{\diamond} \dashv \underset{\omega}{\square}\right) := \left(\left(\omega^* \circ \underset{w \,:\, \omega^{-1}(-)}{\exists}\right) \dashv \left(\omega^* \circ \underset{w \,:\, \omega^{-1}(-)}{\forall}\right)\right) \; : \; \mathbf{H}_{/W} \longrightarrow \mathbf{H}_{/W}.$$

In traditional possible-world semantics such an equivalence relation is called an *accessibility* relation between possible worlds. Now

- $\diamond_\omega p$ is true at $w \in W$ if and only if it is true at at least one $\tilde{w}$ in the same equivalence class as $w$.

- $\square_\omega p$ is true at $w \in W$ if and only if it is true at all $\tilde{w}$ in the same equivalence classes as $w$.

We still find ourselves dealing with an S5 kind of modal type theory. In §4.3 we move away from symmetrically accessible worlds.

Now, even though we have achieved a successful encoding of S5 modalities within the propositional portion of a dependent type theory by relying on variation over a type of worlds, there is still much to be learned from this construction by those who refuse to countenance what they take to be such metaphysical fantasies as possible worlds. The fact that I could begin my account with variation over dogs should indicate to us that what is really at stake is variation broadly construed. A type of worlds can be considered as playing the role of a *generic* domain of variation, rather as the probability theorist employs the notion of a sample space, $\Omega$, as a domain for their random variables.

Consider how for $A$, a type, and $B$, a property of that type, we mark the difference between the following:

- This $A$ is $B$.
- This $A$ is necessarily $B$ (by virtue of being $A$).

Using famous examples from Nelson Goodman's 1955 book *Fact, Fiction, and Forecast*, let us contrast

- This coin in my pocket is silver.
- This emerald is necessarily green.

Where I might look at a gem, $e$, which I know to be an emerald, and observe that it is green, this would only warrant, for some act of perception, $p$,

- This emerald is green.
- $\vdash p : Green(e)$.

If, on the other hand, I have a witness to a universal statement

- $\vdash f : \prod_{x:Emerald} Green(x),$

then I can apply this function to my gem, $e$, to construct $f(e) : Green(e)$. With its focus on the production of a warrant for a proposition, constructive type theory draws our attention to the way warrants have been generated.

We can see why the language of necessity is invoked from our analysis of modality above. Consider the necessity operator corresponding to the type $A$ through its map to **1**. When applied to a dependent proposition, $x : A \vdash B(x) : Prop$, it produces $x : A \vdash \Box_A B(x) :\equiv A^*(\forall_{x:A} B(x)) : Prop$. Now an element of a type is just a map from the unit type **1** to that type. For instance, a particular element, $a$, of the type $A$ corresponds to a map, $a : \mathbf{1} \to A$. This map will generate three maps between propositions, **H**, and all $A$-dependent propositions, **H**/$A$; in particular, the map $a^*$ which sends an $A$-dependent proposition, $B$, to the proposition in the fibre over $a$, or $B(a)$.

Evaluating our type $\Box_A B(x)$ at $a$ in this way results in $\Box_A B(a) \simeq \forall_{x:A} B(x)$. In other words, in the case of the emerald above, when we found $f$ guaranteeing the greenness of all emeralds, or $f : \forall_{x:Emerald} Green(x)$, it transpires that it can be construed at the same time as an element of the equivalent type $\Box_{Emerald} Green(e)$, a type which explicitly represents the necessity of $e$'s greenness qua emerald. We see, then, that one's entitlement to add 'necessarily' to a claim about the possession of some property of an individual in a type depends on the derivation one has of the element witnessing its truth, as displayed in the syntactical form of that element.

However, it can't just be a question of a true universal statement playing this role. For example, we might also have the universal truth:

- All the coins in my pocket now are silver.

Of course, in a sense, necessity is present here too. If all the coins in my pocket are silver, then a choice of such a coin *must* result in a silver one.[5] On the other hand, there's evidently a difference that Goodman was pointing out with this example in that there seems to be no deeper *modal* element in this case, in the following sense. I've only taken a look at some of the coins that will ever be in my pocket: those which happen to be there now. By tomorrow I may have acquired some copper coins. Indeed, give me a copper coin and I could place such a coin there right now. By contrast, I cannot possibly discover or synthesize a non-green emerald.

But note that we have a necessity operator for every type, and indeed for every function between types. When we assess a judgement for its necessity, we must decide what is a relevant 'natural' range of variation. For my coins, a range time-limited to some moment seems unnatural. Then again, I could purchase some trousers and after a brief time seal up its pocket, never to carry more coins in it. The coins that have ever been in my pocket may all happen to be silver. Even so, there seems to be a difference between a type such

---

[5] A point also noted by Brandom (2015, p. 162).

as *Emerald* and one such as *Coin ever in this pocket*, so that we are unlikely to use the latter as a domain. The former has been of human concern for centuries, certainly since the time of the ancient Egyptians; the latter was cooked up for a philosophical puzzle, and we expect that no language will contain a word for such a type. Furthermore, the truth of '*This emerald is green*' is known to be due to the structural properties it shares with all emeralds. We have discovered that it is the chromium-infused beryl $(Be_3Al_2(SiO_3)_6)$ composition which begins to explain its greenness.[6]

Goodman (1955) presented inductive logicians with a range of challenges in making sense of the difference between law-like regularities and those that just happen to occur. The logical empiricists he was addressing were nervous of modal talk, and so hoped to rely merely on syntactical features of general statements. Rather contorted concepts were devised for this, such as Hempel's *purely qualitative predicates*. Instead, along with the inferentialist, we can say that much rests on the web of inferential relations to which specific types belong, which dictates what we expect to change and remain the same as conditions vary over a range. These expectations, naturally enough, change over time. We have a very different understanding today of what kind of measures could lead to a gem changing its colour compared to the expectations of a seventeenth-century alchemist. But at any moment, any user of language in its primary function of empirical description will possess 'the practical capacity to associate with materially good inferences *ranges of counterfactual robustness*', to speak in Brandom's terms (2015, p. 160). It is this capacity that underpins our use of much modal vocabulary, and type discipline gives rise to this capacity. With HoTT this is very plain, since this form of type theory provides the syntax for $(\infty, 1)$-toposes, a structure that persists when forming any slice. Variation over types or functions as involving transport between these slices gives rise to alethic modalities.

We can tell a similar story within mathematics itself to explain why it appears to be possible there to distinguish law-like regularities from *happenstantial* ones, even though all truths of mathematics are taken as necessary in some sense. The case I gave in Corfield (2004) concerns the theorem

- All primes of the form $4n + 1$ are sums of two squares.

Now consider the difference between establishing the proposition

- 13 is the sum of two squares,

first simply by presenting the solution $13 = 2^2 + 3^2$ and second by using the function, $f$, that is devised in a constructive proof of the general result that will send any prime of the form $4n + 1$ to a pair of appropriate numbers. Even if $f(13) = (2, 3)$, yielding the same two numbers to be squared, the difference between the derivations points beyond the potentially happenstantial nature of the particular result to a truth which holds at all points within a significant range of variation.

---

[6] If you're from North America, infusion by vanadium also counts. Note also that we are ignoring the fact that colour is used in counting such beryl gems as emeralds, making this statement *analytic*.

In sum, our modal vocabulary provides us with the means to make explicit our commitments to the behaviour of entities according to their types. The use of a specific type of *worlds* to construct modal operators is a means to portray a most general form of variation. Let's see now how such variation applies not only to dependent propositions, but to dependent types in general.

## 4.2 Towards Modal HoTT

### 4.2.1 General Types

In the section above, I have denoted the expression 'it is necessarily the case that' as $\square$ and applied it directly to propositions. In view of HoTT's understanding that propositions are simply a kind of type, those types at level $-1$ on the $n$-type hierarchy, we should expect there to be a modal form of types from the full hierarchy. In other words, we should look to form $\square A$ for any type, $A$. But before proceeding along these lines, let's reflect on our options. In the philosophical literature, it is not uncommon to hear of modalities as being expressions which qualify the truth of judgements (Garson 2018). Since in the constructive type theory tradition, the judgement $\vdash P$ *true* for a proposition, $P$, became $\vdash p : P$ in HoTT, what should we make of $\vdash P$ *necessarily true*? We can't apply an operator to the '$p : P$' part itself, but we might think to modify the nature of the judgement, perhaps to something like $\vdash_{nec} p : P$. Alternatively, as we were led to do earlier, we modify the symbols for the types, $\vdash p : \square P$. As we will see, these two choices may be reconciled. Judgement in another stricter domain can be reflected back within the original domain, where it will count as having constructed an element of a modified type. For now, however, let us continue with the second of these strategies.

In §4.1 we saw propositions depending on a type of worlds. But it is perfectly possible to extend the application of the modal monad and comonad to *any* such world-dependent types. One of the very pleasant features of a topos, $\mathbf{H}$, is that, if you take any of its objects, say, $A$, then the so-called *slice category*, $\mathbf{H}/A$, is also a topos. The slice category has as objects maps in the topos $f : B \to A$. A morphism in the slice topos to another object, $g : C \to A$, is a map, $h : B \to C$, such that $f = g \circ h$, the triangle commutes. Analogous statements hold true for $(\infty, 1)$-toposes, the categories which model HoTT. Translating this category-theoretic perspective into type-theoretic talk, the equivalent in type theory of the object $f : B \to A$ in $\mathbf{H}/A$ is the dependent type $x : A \vdash B(x) : Type$. Working in a context is equivalent to working in the associated slice of a topos. The empty context corresponds to the object, $\mathbf{1}$, the unit type, and $\mathbf{H}/\mathbf{1} \simeq \mathbf{H}$.

Then for $\mathbf{H}$, an $(\infty, 1)$-topos, and $f : X \to Y$, an arrow in $\mathbf{H}$, we have that so-called *base change*, playing a similar role to that played above by *owner**, induces an adjoint triple between slices:

$$(\sum_{f} \dashv f^* \dashv \prod_{f}) : \mathbf{H}/X \underset{f_*}{\overset{f_!}{\underset{\leftarrow}{\overset{\to}{\underset{\to}{\overset{f^*}{\leftarrow}}}}}} \mathbf{H}/Y.$$

In the case that $f$ is a map from a type to **1**, we recover the dependent sum and dependent product of Chap. 2. So since these latter make up a special case of the construction that covers general functions, $f$, we often call the left and right adjoints in this general case dependent sum, $\sum_f$, and dependent product, $\prod_f$, as well.

This differs from §4.1 in that that now there is no reason for the dependent sum for a dependent proposition to be itself a proposition. As we saw in 2.4, the dependent sum for an $A$-dependent property $B(x)$ gathers together all $As$ which are $B$. To express that there is merely some $A$ which is $B$, we must *truncate* this type to a proposition.[7] Since any construction in HoTT can be interpreted in any $(\infty, 1)$-topos, this is true of the property of types, *isProp*, the property of being a proposition. Interpreted in the slice category $\mathbf{H}/W$ for some type $W$, this property holding for a $W$-dependent type amounts to there being a monomorphism into $W$. If $W$ is a set of worlds, then a world-dependent proposition is a subset of $W$, the proposition corresponding to the worlds in which it holds. This explains the characterization in the modal logic literature of a proposition as the set of possible worlds in which it holds.

Continuing with dependence on a type of worlds, so $X = W$, a set, and $Y = V$, the equivalence classes of accessible worlds, now consider a world-dependent type, $B(w)$, which we take as a *set* for clarity. Then $\Diamond_W B(w)$ is the collection of pairs $\langle w', b \rangle$, with $w'$ accessible to $w$ and $b : B(w')$. A possible $B$ at a world is an actual $B$ at some related world. Meanwhile, $\Box_W B(w)$ is the collection of maps which for each world, $w'$, accessible to $w$, select an element of the respective type, $B(w')$. So a necessary $B$ at a world is a selection of a $B$ from each of its accessible worlds.

There is a natural map, $B(w) \to \Diamond_W B(w)$, which sends $b : B(w)$ to $\langle w, b \rangle : \Diamond_W B(w)$, and one from $\Box_W B(w) \to B(w)$, which evaluates the map $w' \mapsto b(w') : B(w')$ at world $w$ as $b(w)$. While still being a monad and comonad, respectively, $\Diamond_W$ and $\Box_W$ as defined above are no longer idempotent. Consider the case where all worlds are mutually accessible, that is, $V = \mathbf{1}$. Then $\Diamond_W \Diamond_W B(w)$ is composed of elements $\langle w', \langle w'', b \rangle \rangle$, with $b \in B(w'')$. Of course there is a projection from this element to $\langle w'', b \rangle$, so that we have a natural map, $\Diamond_W \Diamond_W B(w) \to \Diamond_W B(w)$. We can similarly find a natural map $\Box_W B(w) \to \Box_W \Box_W B(w)$.

To illustrate with the example of the type of players depending on the type of teams in a league, then taking all teams as related, so working with the map $Team \to \mathbf{1}$,

- The type $\Diamond_{Team} Player(t)$ amounts to the constant type assigning to a given team all of the league's players.

- The type $\Box_{Team} Player(t)$ amounts to the constant type assigning to a given team all choices of one player from each team.

For instance, a rival team's goalkeeper is a *possible* player for your team, and the selection of a captain from each team, or the selection of the tallest player of each team is a *necessary* player.

Let us now put these constructions to use to provide a setting in which an old chestnut of a puzzle makes sense. A standard example of the perils of substitution runs as follows:

---

[7] There is an adjunction between types and propositions. The monad for this adjunction is the truncation operator we treated in §3.1.2, sending a type, $A$, to $||A||$.

- It is necessarily the case that 8 is greater than 7.
- The number of planets is 8.
- It is necessarily the case that the number of planets is greater than 7.

For simplicity, we keep with the case where all worlds are accessible to one another.

We have, of course, that 7 and 8 are elements of $\mathbb{N}$. But we need a world-dependent version of these numbers. So let us form the trivially world-dependent $W^*\mathbb{N} \simeq W \times \mathbb{N}$, which provides a copy of $\mathbb{N}$ at every world. Versions of our numbers, the constant functions $\lambda w.7$ and $\lambda w.8$, are then elements of $\prod_{w:W} W^*\mathbb{N}$. Now we can compare these world-varying numbers with others, such as 'the number of planets in $w$'. From the order relation $>$ on $\mathbb{N}$, we can define a pointwise relation on world-varying numbers, $>_W$.

Since $8 > 7$ is true, $\lambda w.8 >_W \lambda w.7$ is a true proposition at all worlds, hence necessarily true. This is our version of the 'Necessitation rule' that a theorem in the empty context is necessarily true. We also have that 'the number of planets$(a) = \lambda w.8(a) = 8 : \mathbb{N}$'. So we may derive the truth of (the number of planets $>_W \lambda w.7)(a)$, but not that of

$$\prod_{w:W}(\text{the number of planets} >_W \lambda w.7)(w).$$

In other words, we don't have that it is necessary that the number of planets is greater than 7. What appears to confuse people in the puzzle is that when $m$ is not a constant function in $\prod_{w:W} W^*\mathbb{N}$, then it is not equal to the constant function $\lambda w.m(a)$ for some world, $a$.

Now, returning to non-constant world-dependent types, we may wonder whether in general there is any comparability between worlds. If a type is formed by a map from a set $B$ to $W$, that is, as dependent type $w : W \vdash B(w) : Type$, then there are no associated means to identify elements assigned to different worlds. On the other hand, as we saw with the natural numbers, we can form a world-dependent type by base change from a type existing in the empty context. So we might have a general type, $A$, then form the $W$-dependent type $W \times A$, the dependency represented by the first projection to $W$. Here, of course, an element, $\langle w, a \rangle$, in one world corresponds to the counterpart element in another world, $\langle w', a \rangle$, with the same second element.

We might then form a subtype, $P$, of such a constant type. This will come equipped with its projections to $A$ and to $W$. If for some $a$ in $A$, $W \times \{a\} \subseteq P$, so that a copy of $a$ is found in every world, then we have a *section* in $\prod_{w:W} P(w)$, represented by $f(w) = a$ for all $w$. To illustrate this, consider that the range of foodstuffs is constant across worlds, or let's say here *situations*. In each situation a recipe for a beef stew is to be given, the ingredients for each recipe coming from that fixed range of foodstuffs. Then presumably beef will be an ingredient in each case, whereas potatoes may be left out on occasion. So beef is a *necessary* ingredient of a beef stew, whereas potatoes are not. Elements of $A$ here are acting as *rigid designators* in the sense Kripke gives the term in *Naming and Necessity*:

> Let's call something a *rigid designator* if in every possible world it designates the same object, a *nonrigid* or *accidental designator* if that is not the case. (Kripke 1980, p. 48)

The rigid designator 'Nixon' in Kripke's famous example is acting like 'beef' to pick out the same entity in each world in which it exists. Starting from a non-world dependent type of people, *Person*, we form the dependent type of people in a world, $P(w)$, a subtype of $W^*Person(w)$. Then although in our world *Nixon* and *The person who won the United States presidential election of 1970* refer to the same person, only the former is a rigid designator with respect to *Person*.

Of course, one may question such accounts. Does it make any sense to postulate a set of world-transcending people, *Person*, with which to form $W^*Person$, or even a subtype collecting these people and the range of worlds in which each lives? Consider the fictional worlds presented by the novels of Charles Dickens. The Artful Dodger lives in the world of *Oliver Twist* and Uriah Heep in that of *David Copperfield*. There is no chance of any identification of characters between these worlds. We have a straightforward dependent type,

$$x : Dickens\ Novel \vdash Character(x) : Type,$$

with no thought of identifying elements from different fibres. We could still gather together the characters in a non-dependent type using dependent sum, but little is gained by doing so when any element of this collection only inhabits one fictional world.

Someone who takes the possible worlds idea seriously may reply that while this exclusivity may hold for the separate worlds of fiction, they can envisage a collection of worlds, including the actual one, which covers a vast array of fine variations. If there are two worlds differing only as to the timing to the millisecond of the fall of a leaf from a tree in a vast forest, surely then we expect to find that most individuals present in one world also occur in the other. But *counterpart* theorists, such as David Lewis, depart from any account which allows an individual to be in more than one world. In their version of modal semantics, we don't have individuals inhabiting different worlds, but an element of a second world may be the counterpart of an element in this world if in some sense it is sufficiently similar.

Even if the possible world sceptic is not persuaded by either account, nevertheless there's something structurally interesting happening here concerning ways of relating elements between fibres of maps, something very familiar to mathematicians. Let's illustrate this by leaving behind these speculative worlds and confining ourselves to something occurring in our own familiar world.

Let's consider modalities generated by a simple surjective mapping, the map *spec* from the type of animals to the type of species, the one which assigns to each animal its species. Now take the dependent type of $x : Animal \vdash Leg(x) : Type$. Then an element of $\bigcirc_{spec}Leg(x)$ is any leg of a conspecific of $x$, and an element of $\square_{spec}Leg(x)$ is a description of a leg possessed by each conspecific of $x$. In terms of a dog called Fido, a 'possible leg' for Fido is any dog's leg, while a 'necessary leg' is an assignment of a leg to each dog. For the latter, we could take, for instance, 'the last leg to have left the ground', or 'the right foreleg'.

Then $Legs(Fido) \rightarrow \bigcirc_{spec}Legs(Fido)$ is just the inclusion of Fido's legs into all dogs' legs. This is relevant to the discussion in philosophy as to whether possible objects are counted as pertaining to a world. Fido's possible legs pertain to him even if they make reference to other dogs' legs. Similarly, we have a type of possible entities of a kind at a world, members of which make reference to entities of that kind in another world. But we see here that we must maintain type discipline. An element of $\bigcirc_W A(a)$ lives in $a$ in a sense, but as an element of a specific possible type, not as an element of $A(a)$. For instance, when $A$ is a type of concrete objects, an element of $\bigcirc_W A(a)$ has no place in the world $a$. It has a *possible* place at $a$, but this makes reference to the place in the world in which the corresponding element actually lives.

In general, there won't be a map from an animal-dependent type to its $\Box$-version. Think of the dependent type $x : Animal \vdash Offspring(x) : Type$. My indicating one of an animal's offspring gives me no means to pick out an offspring of a conspecific. Indeed, this type will be empty if some conspecific hasn't reproduced yet. So which types do allow a map to the $\Box$-version? Which are, in some sense, *necessary*? Well, certainly those types *pulled back* from ones dependent over species. A *standard* is being provided then to allow comparison across conspecifics.

We can see this as follows. Given the map, $spec : Animal \to Species$, we have

- $s : Species \vdash BodyPart(s) : Type$
- $\vdash front\ right\ leg : BodyPart(Dog)$
- $x : Animal \vdash spec^*BodyPart(x) : Type$

where this last dependent type assigns to an animal the type of body parts of its species. We now have a map from $spec^*BodyPart(x)$ to $\Box_{spec}spec^*BodyPart(x)$. Given an element in $spec^*BodyPart(Fido)$, such as Fido's front right leg, we can name the corresponding body part for Fido's conspecifics, that is, we can form an element of $\Box_{spec}spec^*BodyPart(Fido)$.[8]

An element generated in such a way might be said to refer to an *essential* characteristic of a dog. If I point to the front right leg of a dog and show you another dog, you will probably choose the same leg. It might be holding this paw in the air, so you could have chosen the left rear leg of the second dog who is cocking this now at a lamp post, but it does seem that the same element of the body plan is the most reasonable choice, certainly more reasonable than via a standard relying on time-ordering of lifting actions.

This quality of being able to transfer between the *fibres* of a map is prevalent throughout mathematics. Indeed, there's a direct route from the simple considerations we have just covered to the topics of connections on fibre bundles and of solutions to (formally integrable) partial differential equations. As I will mention briefly in Chap. 5, these latter equations may be seen as recipes which dictate how behaviour carries over from a point to other points in its infinitesimal neighbourhood.

As a technical aside: the *algebras* for the possibility monad, types for which there is a $W$-dependent map, $\bigcirc_W B \to B$, coincide with the *coalgebras* for the necessity comonad, with corresponding map, $B \to \Box_W B$, and are such that there is a natural map, $\sum_W B \to \prod_W B$. Given an element, $b : B(w')$, from some world, we must generate a function, $f : w \mapsto f(w) : B(w)$, such that $f$ picks out the original element, that is, $f(w') = b$. In other words, we must know how to *continue* the function given its value at a single world.

---

[8] Note that we are assuming that we are dealing with a world in which no animal has lost a leg. Alternatively, we might speak of Patch having lost his front right leg, there being an expectation that such a thing should be present.

Before ending this section, I should say a word on the *homotopical* aspect of modal HoTT. Until this point in the chapter, we have only considered types which are propositions or sets which are depending on a base set. But we may climb the hierarchy and apply the constructions above to the case where our types depend not on a set, but on a group. As in the last chapter, we find that for a type, $A$, acted on by a group, $G$, then $\sum_{*:BG} A(*)$ is a type with structure that of the action groupoid, composed of the orbits, sometimes called the *coinvariants*. Now we can *base change* (or *context extend*) back to a $\mathbf{B}G$-dependent type by applying the trivial action of the group. The result should resemble something like *possibility*. The map $A \to \bigcirc_{BG} A$ may be thought of as sending an element $a$ of $A$ to the collection of the images of $a$ identified under the action of $G$.

There would seem to be something here close to what Cassirer (1944) describes in a paper where he brings together Felix Klein's *Erlangen Program* with the findings of the Gestalt school of psychology. Cassirer is claiming that we can find in perceptual activity the seeds of Klein's idea of geometry as the study of invariants of spaces under transformation by smooth groups. For instance, two lines meeting each other at right angles in the Euclidean plane are sent by a Euclidean transformation of the plane to another such pair. Cassirer here takes up the Gestalt theorists' idea of grasping 'intrinsic necessities' when we perceive sensations as variants of each other.

> The 'images' that we receive from objects, the 'impressions' which sensationalism tried to reduce perception to, exhibit no such unity. Each and every one of these images possesses a peculiarity of its own; they are and remain discrete as far as their contents are concerned. But the analysis of perception discloses a formal factor which supersedes this particularity and disparity. Perception unifies and, as it were, concentrates the manifolds of particular images with which we are supplied at every moment . . . Each invariant of perception is . . . a scheme toward which the particular sense-experiences are orientated and with reference to which they are interpreted. (Cassirer 1944, p. 32)

We might say that rather than a type of discrete images, we take them to be related to each other by transformations, grouping them within their orbits. For each of these distinct orbits, we perceive an entity.[9]

Along with dependent sum, we may similarly form the dependent *product*, which in this case is composed of the 'fixed points' of the action, those elements of the set left unchanged by all elements of the group. So $\square_{BG}$ sends a $G$-action to the trivial action of $G$ on fixed points. Then there's a map from the latter to the original $G$-set, the inclusion, corresponding to the map $\square A \to A$. The *necessity* here translates to invariance under the group action.

Expanding this case to have a base type be a general 1-groupoid, $K$, then a dependent type is the assignment of a fibre, $B(k)$, to each element $k$ of $K$, such that moving along any identity between elements of the base results in an equivalence between the corresponding fibres. Two such identity elements in $Id_K(k_1, k_2)$ may result in different counterparts in $B(k_2)$ for a single element of $B(k_1)$, in a way reminiscent of the relative phase shift experienced by

---

[9] The setting for Cassirer's paper is one where *continuous* or *smooth* transformations take place, that is, where topological and Lie groups operate. We will return to his paper in Chap. 5 when we turn to geometry.

identically prepared particles according to the different paths they take between two points in situations which give rise to the so-called *Aharonov–Bohm effect*. Synthetic homotopy theorists are developing a version of *covering spaces* from algebraic topology in terms of such maps from a base space type $A$ to *Set* (Hou and Harper 2018). Think here of sending a point in the base space to the set of points in its fibre. This is to represent the base space as a groupoid, which might rather have been seen as that space's fundamental groupoid. We will need the spatial modalities of Chap. 5 to allow us to maintain this distinction.

Returning to groups, in the relative case, the map $\mathbf{B}G_1 \to \mathbf{B}G_2$, induced by a group homomorphism, $G_1 \to G_2$, provides an adjoint triple between $G_1$-actions and $G_2$-actions, the outer adjoints corresponding to *induced* and *coinduced* actions (see, for instance, Greenlees and May (1995)). It is remarkable to see how the simple ideas we have proposed for presenting modal type theory are intimately related to cutting-edge research in equivariant homotopy theory.

### 4.2.2 First-order Modal Logic and Barcan

The form of modal HoTT we have developed should give us a first indication of what to make of one of the thorniest issues in first-order modal logic: the Barcan formulas. These are named after Ruth Barcan who sought to compare formulas of the form $\bigcirc \exists x P(x)$ and $\exists x \bigcirc P(x)$, designating claims concerning the possible existence of a $P$ and the existence of a possible $P$. It is perhaps clear why one might wish to resist the *forward* Barcan inference, since it seems to propose that there possibly being something with a property entails that there is something which possibly possesses the property. But how can a mere possibility of existence be enough to entail existence?

Here again, we will need to be careful with the typing discipline. If the modalities apply in the context of a type of worlds, $W$, or equivalently to the slice over $W$, then there is a problem in trying to make them interact with quantification over a world-dependent type. Say we have a world-dependent type, $w : W \vdash B(w) : Type$, and then some further property, $w : W, b : B(w) \vdash P(w, b) : Prop$, then we may quantify over $B$, thereby removing it from the context, and then be able to apply the modal operators. But in this case it won't make sense to apply the modal operators first—they can only apply to the bare context $W$. The only way this reversal could take place is if $B$ is not really $W$-dependent, so that the context is $W, B$. Then we could apply modalities to remain in the $W, B$ context, and then apply quantifiers to land in dependency over $W$.

Indeed, given the constant world-dependent type $W \times B$, derived from a plain type, $B$, we might have a dependent proposition as follows:

$$w : W, x : B \vdash P(w, x) : Prop.$$

Then $w : W \vdash \exists_{x:B} P(w, x) : Prop$, so $w : W \vdash \bigcirc_W \exists_{x:B} P(w, x) : Prop$, in which elements at a world, $a$, will be $(w, (b, p))$, $p$ witnessing that $b$ is $P$ in world $w$. Treating the context now as symmetric, we can also form $w : W, x : B \vdash \bigcirc_W B(w, x) : Prop$ and $w : W \vdash \exists_{x:B} \bigcirc_W B(w, x) : Prop$. Here I'm not quantifying over things in my world, but rather over world-independent $B$. The Barcan formula is saying that some member of $B$ turning out to have $P$

when it appears in some world is equivalent to there being some world containing a $B$ which is $P$. Along the Kripkean line of transworld individuals, these are evidently the same claim.

In general, however, the type $B$ may genuinely depend on $W$. Then we haven't a means to exchange as above. But say I have, as above, $w : W, x : B(w) \vdash P(w,x) : Prop$. Then I can form $\bigcirc_W \sum_{x:B(w)} P(w,x)$, the world-dependent (constant) type containing all possible $B$s which are $P$. Evaluated at my world $a$, this is $\sum_{w:W} \sum_{x:B(w)} P(w,x)$. I can also form $\sum_{(w,x):\sum_{w:W} B(w)} P(w,x)$. Then an interpretation of the Barcan formulas amounts to the equivalence of taking dependent sum in one or two stages:

$$(\bigcirc_W \sum_{x:B(w)} P(w,x))(a) \simeq \sum_{w:W} \sum_{x:B(w)} P(w,x) \simeq \sum_{(w,x):\sum_{w:W} B(w)} P(w,x).$$

The second equivalence is just the rebracketing of the three-part terms in pairs: $(w,(b,p)) \leftrightarrow ((w,b),p)$. We might say:

- At this world, there is possibly a $B$ which is $P$.
- At this world, there is a possible $B$ which is $P$.

This solution goes in some respects along the lines of Timothy Williamson (2013) to allow quantification over possible things. However, we arrive at this solution maintaining strict typing discipline, and certainly not allowing untyped quantification over 'everything'. Modal dependent type theory would appear to provide a helpful system of constraints guiding reasonable forms of expression.[10]

### 4.2.3 Contexts and Counterfactuals

Of the many philosophers who do not go along with the Barcan formula, let us briefly consider an account by Reina Hayaki (2003), who claims that there is no proper sense in which merely possible entities exist.

All apparent references to non-actual objects are circumlocutions either for *de dicto* statements about ways the world might have been, or for *de re* statements about ways that actual objects might have been, or for some combination of both. (2003, p. 150)

Since, for her, possible objects do not exist in our world, she needs to make sense of what appear to be truths we can utter about possible objects. Consider the following pair of propositions relating to the example she discusses:

I could have had an elder brother who was a banker. He could have been a concert pianist if he had practised harder.

---

[10] The resources we have developed up to this point could be used to explore what is studied in so-called *two-dimensional semantics*. This concerns the use of actuality or indexicals, as with 'It is possible for everything that is actually red to be shiny', to be rendered perhaps as 'There is some world, $b$, in which everything that is red in this actual world, $a$, is shiny there in $b$'.

It appears to be the case here that we are referring to a non-existent being, an elder brother, to say counterfactual things about him. If we restrict ourselves to worlds arranged on an equal footing, it certainly sounds as though we are treating this possible man *de re*.

However Hayaki explains the situation through nested trees, where the first sentence presents a level 1 world, and the second a level 2 world. Our world might have gone differently with my parents having another son whose career was in banking. Then in that level 1 world, of the object that is that man it can be said that he might have had a different profession. Hayaki considers this more highly motivated version of her imagined brother as inhabiting a level 2 world.

What I want to take from this construction is the idea that there's a structure to the variability of possibility that goes beyond variation over a set. Hayaki talks in terms of stories and their continuation. Let's follow her in this by considering winding back the story to a time before her birth but after her parents met, and then winding forward again with their two children and career choices. Leaving aside whether there would be a 'she' if her parents had had a child before her, we can spin out two tales from the assumption of that child's birth and hers, according to the two choices of profession. Then, once we have a branch with a banker brother, we can speak of his possibly being a pianist, since we need only wind that branch back to a point where career decisions are being made. But rather than this informal talk of stories and continuations, let's see if they can be included with the formalism of the modal calculus we have been developing.

In this chapter we have been considering the type of worlds as the space of variation for our modal considerations, but given the role of contexts in HoTT as providing the typed variables for terms and further types, we might explore whether contexts themselves could act as a way to formulate worlds. We can see this idea of winding back through history in terms of deconstructing the type of worlds. Indeed, recall Ranta's idea from §2.5 that a context is like a narrative, where we build up a series of assertions, any one of which may depend upon previously introduced terms.

> A man enters a saloon. He is whistling *Yankie Doodle*. A woman enters the same saloon, holding the hand of her child. It is his wife.

As before, think of a work of fiction introducing the reader to things, people and places, describing their features, their actions, and so on. Then possible worlds relative to what has been stated so far are ways of continuing: new kinds of thing may be introduced, or new examples of existing kinds, and identities may be formed, such as when we find out that Oliver Twist's kindly gentleman is in fact his grandfather. And so on. Necessity then describes what must happen, and possibility what may happen. So for our story we could ask:

> Must the child be the man's daughter? Might she call out to him? Must they all be in the same room?

These are all questions about ways the story may or will continue. We might also wonder whether the story could have gone differently:

Might he have been humming rather than whistling? Might he have whistled a symphony? Might the woman have held the child in her arms?

Now, formally, recall that a context has the form:

$$\Gamma = x_0 : A_0, x_1 : A_1(x_0), x_2 : A_2(x_0, x_1), \ldots, x_n : A_n(x_0, \ldots, x_{n-1}).$$

In view of the pleasant category-theoretic setting of HoTT, any such context corresponds itself to an object, the iterated dependent sum of the context. Let $W$ represent the iterated sum, and $W_i$ the stages of the construction of $W$, then the maps we considered earlier,

$$\mathbf{H}_{/W} \underset{\longleftarrow}{\overset{\longrightarrow}{\longrightarrow}} \mathbf{H},$$

now factor through the successive stages of the construction of the context:

$$\mathbf{H}_{/W} \underset{\longleftarrow}{\overset{\longrightarrow}{\longrightarrow}} \cdots \underset{\longleftarrow}{\overset{\longrightarrow}{\longrightarrow}} \mathbf{H}_{/W_2} \underset{\longleftarrow}{\overset{\longrightarrow}{\longrightarrow}} \mathbf{H}_{/W_1} \underset{\longleftarrow}{\overset{\longrightarrow}{\longrightarrow}} \mathbf{H}.$$

Still there are different ways as to how to take this idea further.

There is such a vast store in our shared context that it seems that a story can go almost anywhere its author wishes. Continuing our Western, an elephant escapes from the box car in which it travels with the circus and tramples all in the saloon underfoot. Or, a tornado rips through the town and takes the child somewhere over the rainbow. Instead, one might imagine a more controlled setting of what can occur next, where paths fan out according to circumscribed choices, as we find with computation paths in computer science or, in a more extreme form, with the collection of real numbers for the intuitionist. Ranta (1991) began the exploration of these themes from the perspective of dependent type theory. Like the real numbers formed from all possible infinite decimal expansions, here we can conceive of the specification of a collection of worlds which are all possible 'complete' extensions of a context, $\Gamma$.

> Worlds appear as total infinite *extensions* of finitely representable *approximations* of them. Moreover, all we can say about a world is on the basis of some finite approximation of it, and hence at the same time about indefinitely many worlds extending that approximation. (Ranta 1991, p. 79)

Where the real numbers enjoy the property that we are simply making the choice of a digit from a fixed set at each turn, in the case of narratives we could think to make a circumscribed story by selecting some characters out of a list of possibilities, specifying some of their possible properties, specifying possible relations between them, specifying possible actions that one character does to another and some consequences of these actions, and so on.

This context-based formulation may give us a way to think about the nature of counterfactuals, such as

- Had I taken an aspirin this morning, I wouldn't have a headache now.

To make sense of the truth of such counterfactual statements, David Lewis famously invoked the notion of a 'closest' possible world in which the antecedent holds. Here we don't have a metric on our space of worlds, but in that we see the collection of contexts as forming a tree, where an extension shifts along a branch, then one could imagine some kind of minimal stripping back of context to leave out that part which conflicts with the antecedent of the counterfactual, before building up to a context where it holds. Then contexts whose shared initial stage is longer will count as closer in something like a *tree metric*.

But we don't want to count all extensions from the common initial stage as equally close. In the case of my headache, specifying a world by turning back to this morning, asserting that I do take an aspirin, but also that I hit my head on a low ceiling, so that I do in that case have a headache, this should count as being at a further distance than a situation where things continued as similarly as possible. We might then want to think harder about the dependency structure of a context. Although a context is given in general as

$$\Gamma = x_0 : A_0, \ x_1 : A_1(x_0), \ x_2 : A_2(x_0, x_1), \ \dots, x_n : A_n(x_0, \dots, x_{n-1}),$$

$A_2$, say, might only really depend on only *one* of its predecessors. The direct dependency graph between types in a context is a *directed acyclic graph*, or a *DAG*. These are famous for expressing the dependency structure of Bayesian networks, a way of representing probability distributions based on causal dependencies. They were developed greatly by Judea Pearl, and described in his book *Causality* (Pearl 2009), where one thing he uses them for is counterfactual reasoning. One minimally modifies the network compatibly with the counterfactual information. We could similarly imagine minimal modifications of context here.

With the idea of branching histories we have come very close to the variants of temporal logic known as computational tree logics. Let us now see if we can develop a temporal type theory.

## 4.3 Temporal Type Theory

The philosophical logic literature makes play of the similarity between temporal modalities and those of necessity and possibility, although now with two pairs, oriented according to the time direction. For example, corresponding to possibility, for $\phi$, a proposition, we have

- $F\phi$ is '$\phi$ will be true at some future time'.
- $P\phi$ is '$\phi$ was true at some past time'.

Just as with classical treatments of modal logic where possibility and necessity are interdefinable, logicians then look to form dual modalities of $F$ and $P$ as follows:

- The dual of $F$ is $G$, so $G\phi = \neg F \neg \phi$. This means that we read $G\phi$ as 'at no future time is $\phi$ not true', that is, $\phi$ is always going to be true. ($G$ is for 'Going'.)
- The dual of $P$ is written $H$, whence $H\phi = \neg P \neg \phi$ and $H\phi$ interprets as '$\phi$ has always been true'. ($H$ is for 'Has'.)

In view of the fact that we could construct the dual modalities possibly–necessarily without negation, we should expect a similar treatment to be feasible here.

When in the treatment of possible worlds we used the existence of a map, $f : W \to V$, we were understanding worlds in the same preimage of $f$ as related or accessible to one another. But another way to view an equivalence relation, and indeed a binary relation more generally, is as a subcollection of the cartesian product of the collection of worlds with itself. So

$$R \hookrightarrow W^2$$

is the collection of related pairs of worlds. Naturally, from $R$ we have two projections, $p, q$ to $W$, to the first and second members of these pairs. We could then, instead of deploying the map $f$, generate our modal operations on the slice over $W$ using $\sum_p q^*$ for 'possibly' and $\prod_p q^*$ for 'necessarily'. When we are dealing with an equivalence relation, $R$, switching $p$ and $q$ in these operations won't lead to anything new. It is worth exploring, then, what would result from a more general relation.

Temporal logicians have long debated the relevant advantages of instant-based and interval-based approaches. Some have also considered hybrid approaches (Balbiani et al. 2011). As we shall see, the analysis of this section suggests that working with intervals and instants together in the form of something like what is called an *internal category* allows for a natural treatment via adjunctions. Indeed, intervals may be construed as given by pairs of instants marking their boundary, and so as

$$Interval \hookrightarrow Instant^2,$$

where the first instant precedes the second. So consider a category, **H**, and an internal relation given by $b, e : Time_1 \rightrightarrows Time_0$. Here we understand elements of $Time_1$ as time intervals, and $b$ and $e$ as marking their beginning and end points. Now each arrow, $b$ and $e$, generates an adjoint triple, for example, $\sum_b \dashv b^* \dashv \prod_b$, formed of dependent sum, base change, dependent product, going between the slices $\mathbf{H}/Time_1$ and $\mathbf{H}/Time_0$.

Then along with the monads and comonads generated by composition within a triple, we can also construct some across the triples. Specifically, we find two adjunctions, $\sum_b e^* \dashv \prod_e b^*$ and $\sum_e b^* \dashv \prod_b e^*$. This results in isomorphisms such as

$$Hom(\sum_b e^* C(t), D(t)) = Hom(e^* C(t), b^* D(t)) = Hom(C(t), \prod_e b^* D(t)).$$

Now consider for the moment that $C$ and $D$ are propositions depending on time instants. Then $\sum_b e^* C(t)$ will contain all instances of intervals beginning at time $t$ where $C$ is true at the end. If this type is inhabited it means 'there is some interval beginning now and such that $C$ is true at its end', that is, $FC$, or '$C$ will be the case'. On the other hand, $\prod_e b^* D(t)$ means 'for all intervals ending at $t$, $D$ is true at their beginning', that is, $HD$, or '$D$ has always been the case'. Hence our adjunction is $F \dashv H$. Similarly, interchanging $b$ and $e$, we find $P \dashv G$. Note that we did not have to assume the classical principle $G\phi = \neg F\neg\phi$ and $H\phi = \neg P\neg\phi$.[11]

---

[11] Similar constructions for relations representing parts of a common whole in 1-toposes are given in Fong et al. (2018), where the authors speak of *inter-modalities*.

Whenever we have a monad, $M$, on a category, $\mathcal{C}$, there is an associated *unit*. This is a map which will send any object of the category, $A$, to its image, $M(A)$. In the case of the free monoid monad described in §4.1, recall that the unit corresponds to a map sending an element of a set to the associated singleton string. Dually, for a comonad, $T$, there is a *counit* from $T(A)$ to $A$. These two maps correspond in the modal cases discussed earlier to the maps, $A \rightarrow \bigcirc_W A$ and $\square_W A \rightarrow A$. We can also see that these constructions are familiar in the adjunctions giving rise to product and sum from §2.2,

- $Hom_{\mathcal{C} \times \mathcal{C}}((A, A), (B, C)) \cong Hom_{\mathcal{C}}(A, B \times C)$.
- $Hom_{\mathcal{C} \times \mathcal{C}}((B, C), (A, A)) \cong Hom_{\mathcal{C}}(B + C, A)$.

The counit for the comonad generated by the former adjunction is a map in

$$Hom_{\mathcal{C} \times \mathcal{C}}((A \times B, A \times B), (A, B))$$

which corresponds to the pair of *elimination* rules for product, and so conjunction. The unit for the monad generated by the latter adjunction is a map in

$$Hom_{\mathcal{C} \times \mathcal{C}}((A, B), (A + B, A + B))$$

which corresponds to the pair of *introduction* rules for sum, and so disjunction.[12]

Now since in the temporal situation we again have monads and comonads, we can consider the various units and counits. These are

- $\phi \rightarrow GP\phi$: 'What is, will always have been.'
- $PG\phi \rightarrow \phi$: 'What came to be always so, is.'
- $\phi \rightarrow HF\phi$: 'What is, has always been to come.'
- $FH\phi \rightarrow \phi$: 'What always will have been, is.'

As before, in the setting of dependent type theory, we do not need to restrict to propositions, but can treat the temporal operators on general time-dependent types. So if *People*$(t)$ is the type of people alive at $t$, *FPeople*$(t)$ is the type of people alive at a point in the future of $t$ and *GPeople*$(t)$ is a function from future times to people alive at that time. For instance, an element of this latter time is 'The oldest person alive$(t)$', assuming humanity continues.

We can then think of adding other features, such as insisting that *Time* be an internal category, and so requiring there to be a composition between any two intervals, the end of one matching the beginning of the other. We may also choose to impose additional structure, such as that the internal category be an internal poset, or a linear order. Let's consider here *Time* as a category, where we have in addition to the two projections from pairs of intervals

---

[12] The other introduction and elimination rules also arise from the adjunctions. The unit for the adjunction producing exponential objects at the end of §2.2 is the application map $C^B \times B \rightarrow C$, which in the case of propositions is the elimination rule for implication.

that adjoin, $p, q : Time_1 \times_{Time_0} Time_1 \to Time_1$, a composition, $c : Time_1 \times_{Time_0} Time_1 \to Time_1$. This allows us to represent more subtle temporal expressions. We could define a property of time instants that a lightning strike happen at that moment, $t : Time_0 \vdash L(t) : Type$. Then we could characterize the property of an interval that it contains a lightning strike as $\sum_c (ep)^* L$ (note $ep = bq$).

We can also represent *since* and *until*. '$\phi$ has been true since a time when $\psi$ was true', denoted $\phi S \psi$ in the literature, is represented as:

$$\phi S \psi := \Sigma_e (b^* \psi \times \Pi_c (ep)^* \phi).$$

That is, there is an interval ending now such that $\psi$ was true at its beginning and $\phi$ was true at all points inside it. Similarly, '$\phi$ will be true until a time when $\psi$ is true' is

$$\phi U \psi := \Sigma_b (e^* \psi \times \Pi_c (ep)^* \phi).$$

To be precise, this last type is such that any inhabitant of it tells us that there is an interval beginning now such that $\phi$ holds at each of its points, and $\psi$ holds at the end. Of course, $\phi$ may continue to hold after the end of this interval. We could easily express variants where $\phi$ no longer holds after $\psi$ first occurs, or allow the use of 'until' in the sense where the condition $\psi$ may never happen.

Properties may also be defined as pertaining to intervals. Then we could express of a given interval that $\alpha$ holds for an initial part of it, and $\beta$ holds for the other part. This has been studied (Venema 1991) in the guise of what is called the *chop* operator, $C$, so then $\alpha C \beta$ denotes this composite property of intervals, as in 'dinner = starter $C$ main $C$ dessert'. In our current framework, we have $\alpha C \beta :\equiv \Sigma_c (p^* \alpha \times q^* \beta)$.

There is also the instant interval map, $i : Time_0 \to Time_1$, which enables us to send a property of intervals, $P(t_1, t_2)$, to a property of times by seeing whether that property holds of the relevant instant interval, $P'(t) := P([t, t])$. Note that this is different from the evaluation of a varying quantity at some moment. Say we have $t : Time_0 \vdash f(t) : \mathbb{R}$, then of course $f(a) : \mathbb{R}$ at some moment, $a : Time_0$, and we may then form a proposition concerning this instantaneous value. So we should agree with the following:

> *Instantaneous events* are represented by time intervals and should be distinguished from *instantaneous holding of fluents*, which are evaluated at time points. Formally, the point $a$ should be distinguished from the interval $[a, a]$ and the truths in these should not necessarily imply each other. (Balbiani et al. 2011, p. 32)

Note that one of the consequences of taking *Time* as an internal category is that the future includes the present, so that $\phi$ could be true now and at no other instant, but we would have that $F\phi$ is true when we may imagine that it is supposed to say '$\phi$ will be true at some Future time'. Similarly, we would have that $\phi U \psi$ holds now if $\psi$ and $\phi$ both hold now (in general, as defined above it requires $\phi$ to still hold at the instant when $\psi$ becomes true). If we wish to change these consequences, we could let $Time_1$ collect the $<$-intervals instead of the $\leq$-ones. In other words, we could take *Time* to be a semicategory. While this accords with standard practice, the original alternative has been proposed:

> The most common practice in temporal logic is to regard time as an irreflexive ordering, so that 'henceforth', meaning 'at all future times', does not refer to the present moment. On the other hand, the Greek philosopher Diodorus proposed that the necessary be identified with that which is now and will always be the case. This suggests a temporal interpretation of □ that is naturally formalised by using reflexive orderings. (Goldblatt 1992, p. 44)

On the other hand, some temporal logicians look to represent both forms of 'henceforth'.

There are many other decisions to be made in modelling *Time*: linear versus branching, discrete versus continuous, dense or not dense, bounded or unbounded, deterministic or undeterministic, and so on. Computer scientists have formulated various calculi to represent these choices. For instance, branching behaviour is captured in $CLT^*$, a computational tree logic. This calculus allows for quantification between branches as well as along branches, so that one might say of a given state 'It is always going to be that the machine will reach state $S$', or 'It is possibly going to be that henceforth the machine is in state $S$'. Such tree logics are used in chip design and verification, as is explained well in Halpern et al. (2001). A type-theoretic version should be easy to formulate, and could very well be useful here.

With some ideas on a temporal type theory in place, let us see if we can make sense of one of Bede Rundle's counterexamples to '*and*' being treated as mere conjunction from §2.3:

- Alice used to lie in the sun and play cards.

For this proposition to be true, it appears that we need several inhabitants in the dependent sum of past time intervals during which Alice lies in the sun and which contain subintervals in which she plays cards. We need the terminal points of the intervals and subintervals to mark the beginnings and ends of the respective activities. Individuation of playing and lying-in-the-sun activities as events will then rely on a number of things, including their timing:

> a necessary condition for the identity of events is that they take place over exactly the same period of time. (Davidson 2001, p. 124)

So we have a map from *Activity* to $Time_1$, which generates the dependent type of activities lasting over an interval, $i : Time_1 \vdash Activity(i) : Type$, and we must only have identity of activities as occurring in the same fibre.

It would appear that we are fast approaching the material on *event nuclei* from §2.6. Recall from there that an event nucleus is composed of a preparatory activity, culminating in an achievement, resulting in a change of state. This neatly matches our set-up. If an event nucleus takes place over an interval, some subdivision of it via the chop operator into two adjoining subintervals and the instant, or, perhaps better, momentary interval where they abut, corresponds to the timing of its parts. So '*He reached for the switch and lit up the room*', covers the preparatory motion to the switch, culminating in its being flipped, resulting in light shining there. Such an event nucleus would be an element of the following type:

$$\sum_{i : Time_1} \sum_{(j,k) : c^*(i)} reach\text{-}switch(j) \times light\text{-}room(k) \times flip\text{-}switch([ej, bk]),$$

where much remains to be specified about these components as activities, achievements and changes of state. Whether we use the perfect rather than the simple past when describing such an episode rests on whether the time interval of the changed state includes the present moment of utterance. We will not say '*He has lit up the room*' if someone shortly after his action flipped the switch off.

Variations are possible. Perhaps one would like the achievement to take place in an instant, rather than a brief interval. Linguists have wrestled with such choices informally:

> theories differ as to whether they take intervals as the basic temporal primitive, and regard events as durative, or whether they take instants as primitive and intervals as composite. Under the first view, a Vendlerian Activity like *running* would be represented as a transition, with a temporal and spatial extent. Under the second view, an Activity would be regarded as a progressive fluent, or property of a state, with the states that it characterizes being accessed via instantaneous incipitative events of *beginning running* and abandoned via terminative events of *stopping running*. (Vendler and his followers seem equivocal between these two interpretations.) Under the latter interpretation, the instantaneous incipitative and terminative events themselves correspond to Vendlerian Achievements, associated with further changes in fluents corresponding to consequent states, such as *running* and *having stopped running*. Vendlerian Accomplishments like *running to the bus stop* are then the composition of an Activity of *running with the goal of being at the bus stop*, the terminative Achievement of *stopping running* and the culminative achievement of *reaching the bus stop*, which in turn initiates its own consequent state of *being at the bus stop*. (Steedman 2012, p. 110)

There is plenty to do here, including making comparisons to other related approaches to temporal types: for instance, the book length treatment (Schultz and Spivak 2017). Something we might do to put their relative power to the test is to take up the challenge to represent complex pieces of script. Balbiani et al. (2011) propose the following examples:

- Ever since he met her for the first time, he could not stop thinking about her and kept calling her several times every night until she would give him a brush-off, and then after being silent for a while he would phone again.
- At the exact moment in which the train passes over the sensor, the rail crossing bar starts to close; the bar will start to open again a while after the train passes over the second sensor.

I think I have shown that the ingredients are available to represent such statements, and others such as the one from Chap. 2:

- It took me two days to learn to play the Minute Waltz in sixty seconds for more than an hour.

But now I want to turn to a very recent effort to place modal type theory in a powerful general framework.

# 4.4 Mode Theory

I will end this chapter with a brief discussion of a very interesting body of work which is currently unfolding. Recall the discussion in §4.2 above, where I offered two options as to how to modify $\vdash p : P$ with a modality. The one that we went on to use was $\vdash p : \Box P$; the other was to tag the turnstile sign, $\vdash_{nec} p : P$. Let us see now if we can make sense of this latter approach.

The idea here is that we have different arenas in which reasoning can take place. In these arenas, different rules may apply as to the inferences permissible there. Then even though these inference rules vary, it is possible for communication to take place, or rather representation of another's reasoning in one's own terms, as we saw in the discussion of how the dog and person experts communicate in the first section of this chapter. Something along these lines was suggested by Haskell Curry, as Fairtlough and Mendler explain:

> Curry's proposal was to take $\bigcirc \phi$ as the statement 'in some stronger (outer) theory, $\phi$ holds'. As examples of such nested systems of reasoning (with two levels) he suggested Mathematics as the inner and Physics as the outer system, or Physics as the inner system and Biology as the Outer. In both examples the outer system is more encompassing than the inner system where reasoning follows a more rigid notion of truth and deduction. The modality $\bigcirc$, which Curry conceived of as a modality of possibility, is a way of reflecting the relaxed, outer notion of truth within the inner system. (Fairtlough and Mendler 2002, p. 66)

We can illustrate this account in terms of the real suggestions of mathematical physicists Jaffe and Quinn (1993), who, alarmed at the lack of rigour they saw to be intruding into their field, proposed to have less rigorous physicist-style arguments for a mathematical result marked as '*theoretical*'. We might say that if physicists have argued something to their own satisfaction, and mathematicians have not decided either way, then the latter should say that it is *possibly* the case. Similarly, with roles reversed, if the mathematician has proved a result for the physicist, or the physicist for the biologist, it should be marked as *necessarily* true.

Fairtlough and Mendler continue in their article by examining whether inference in the outer system might be represented by operators of the form

$$\bigcirc_K^L \phi \equiv K \supset (\phi \vee L),$$

where we think of $K$ as expressing additional resources for reasoning in that system, and where we do not need to establish $\phi$ if we can show that $L$ holds. Reasoners in the outer system have advantages over those in the inner system. They have more resources to deploy and they may find a condition obtains meaning that no further work towards the original conjecture is needed.

Now, of course there's a question of what is meant by speaking of the *same* proposition in different systems. To the extent that one takes the meaning of a proposition to be determined in part by the inference rules present, taking one across verbatim to another setting should alter its meaning. Indeed, this is so—a correct formalism needs to mark this passage. There may be instances where one category of inference is a subcategory of another, and it is easy there to slip into the practice of *coercing* the members of the subcategory to count as members of the full category, rather as one coerces a rational number to count as a real number. However, the general situation requires marking of translation between settings.[13]

Very recent work[14] is pursuing this line of thinking in terms of a modal type theory in which a theory of the relevant modes provides one level of syntax on which can be built the reasoning pertaining to that mode. The deepest level of syntax specifies modes, associated to each of which is a class of types. Arrows between modes, say, $\alpha : p \rightarrow q$, correspond to adjunctions between these classes of types. Then we may have sequents of the form $A_p \vdash_\alpha B_q$.

In a sense we have already seen this mediation between arenas of inference when we took up the triple adjunction between slices,

$$(\sum_f \dashv f^* \dashv \prod_f) : \mathbf{H}/X \overset{\overset{f_!}{\rightarrow}}{\underset{\underset{f_*}{\rightarrow}}{\overset{f^*}{\leftarrow}}} \mathbf{H}/Y.$$

The *modes* here are variation over $X$ and variation over $Y$. They generate a pair of left and right adjoint couples, otherwise known as geometric morphisms. Now, as applied to the case of modal HoTT, Licata et al. generalize this situation so as to take as the basic entity a single geometric morphism, that is, an adjoint pair, between any two $(\infty, 1)$-toposes, with no requirement that they be slices of a common $(\infty, 1)$-topos. Then there may be multiple such mode morphisms between the various modes. Previous attempts had restricted to a partial order of modes, so at most one adjunction between any pair of modes.

This project has a very expressive scope and is looking to provide a syntactical framework for a wide range of modal type theories, including modal HoTT. The slogan here is that, where HoTT itself is the internal language of $(\infty, 1)$-toposes, *modal* HoTT is the internal language for collections of $(\infty, 1)$-toposes related by geometric morphisms. In its full generality, this approach to versions of modal type theory is making sense of a range of previous attempts, and fits smoothly with the relevant mathematics. Harper's *computational* trinitarianism has become *homotopical* trinitarianism (Shulman 2018b). As yet, there is no syntactical formulation which picks out those modal type theories which are only to be understood as describing the passage between slices of categories. Modes which involve *variation over a type* or *over a function* make up only a portion of all mode theories, even if we can use such variation in devising models for modal theories, such as when Awodey and Kishida (2012) employ sheaf models to demonstrate their completeness for first-order modal logic. Such variation relies on the cohesiveness of topological spaces, and this cohesiveness itself turns out to involve spatial modalities which are not to be construed as

---

[13] As Jason Reed claims, 'it is helpful to live in a world where the sort of thing that is eligible to be true is different from the sort of thing that is eligible to be, for instance, necessarily true. In a slogan: different judgements judge different things.' (Reed 2009, p. 1).

[14] See Licata and Shulman (2016), Licata et al. (2017).

variation over types, as we shall see in the following chapter. Another example comes from linear logic, a logic that does not contain the rules for *contraction* or *weakening*, so that it matters precisely how many times a premise appears in a piece of inference.[15] One associated modality, called *Of course* and denoted !, converts a premise, *P*, to be used once in a piece of inference into a premise, !*P*, equivalent to making *P* available for use as many times as desired. In the form of a linear *dependent type theory*, we can understand its semantics as involving an adjunction between a linear and a nonlinear setting.

If, as I argued in Chap. 2, natural language relies on constructions in dependent type theory, we should expect impressive achievements from its integration with monadic and comonadic adjunctions. We should be able to use this formalism to make common cause with Brandom's *pragmatic expressivism*. Integrating dependent type theory, and more specifically HoTT, with the adjunctions generating the monads of computational effects and comonads of coeffects will allow enormous expressiveness, both in computer science and natural language semantics. Already pragmatic aspects of speech are being represented in terms of extensions of simple type theories by monads:

> Side effects are to programming languages what pragmatics are to natural languages: they both study how expressions interact with the worlds of their users. It might then come as no surprise that phenomena such as anaphora, presupposition, deixis and conventional implicature yield a monadic description. (Maršík and Amblard 2016, p. 259)

Indeed, indexical aspects of assertion, such as place and time, and the identity of the asserter are involved here. Concerning the latter, very interesting work by Asudeh and Giorgolo (2016) treats referential opacity using monads arising from the kind of adjunction we treated in §4.1. Now we consider a set of indices, *I*, to correspond not to a collection of worlds, but instead to a collection of individuals. Rather than compose the adjunctions to form a monad and comonad on the slice **H**/*I*, where **H** is the category of our inferences, we compose in the opposite order to form a monad and comonad on **H**. In computer science these go by the names of the *reader monad* and the *writer comonad*. The former has the simple effect of turning a type, say, *A*, into the type [*I*, *A*]. With a term of the latter type, before we can arrive at a term of type *A* we need to supply (or *read* from the situation) a term of type *I*. We may parse its elements not as terms in *A*, but rather as terms denoting elements of *A according to* the different *i* in *I*. This allows for the representation of both *de re* and *de dicto* ascriptions, as when we say to someone '*Your "mother" is your father*' in a situation where we know the person downstairs to be their father, and yet they believe it to be their mother. We may also express person *x*'s perspective on person *y*'s perspective on the reference of a term with the iteration provided by the reader monad, [*I*, [*I*, *A*]]. The structure of this monad allows us to reduce *x*'s perspective on *x*'s perspective to just *x*'s perspective.

---

[15] This gives us yet more subtle ways to treat '*and*', so that we may distinguish in '*You can afford to buy A and you can afford to buy B*' between the reading that allows both purchases to be made simultaneously and the reading that only guarantees you can buy either but not both.

Asudeh and Giogolo also treat the phenomenon of the non-substitutability of corefer-
ential terms. For instance, we may be able to say that Lois Lane loves Clark Kent, and that
he and Superman are the same person, without being able to say that she loves Superman.
The intentional aspect of love requires the lover to have identified the two names as referring
to the same person if we are to assert this affection to be maintained across the identity.

In doing this kind of work, these authors are not deploying their extension of the simply
typed lambda calculus by monads merely to *represent* interesting linguistic phenomena.
Rather, there is a practical concern with providing a computational linguistics for these
'*perspectival*' situations. If computational expense were of no concern and mere representability in principle all that mattered, one might take *any* term of our language to have
this perspectival aspect. What their use of the reader monad allows, however, is the *effective*
integration of perspectival and absolute terms. It would be interesting then to explore what a
monadic extension of a full dependent type theory might enable, as the latest work of Licata
et al. sets out to achieve.

Finally, for the pragmatist, this emerging mode theory may provide a useful opportunity
to revisit Charles Peirce's *gamma* system of *existential graphs*. Peirce thought very highly
of his work on these graphs. The *alpha* system corresponds to a propositional logic, while
the *beta* system corresponds to a kind of first-order logic. These systems have been translated into a category-theoretic framework by Brady and Trimble (2000a, 2000b). We see
varieties of this diagrammatic reasoning calculus deployed elsewhere (for example, Melliès
and Zeilberger 2016). The *gamma* system is generally interpreted as Peirce's attempt to
formulate a modal logic. The system was far from finished, and had gone through various
phases by the time he discussed a late version in 'Prolegomena to an Apology for Pragmaticism' (Peirce 1906). Assertions of propositions on coloured sheets, chosen by Peirce
in accordance with heraldic tinctures (jules, azure, argent, and so on), four tinctures per
class (possibility—Colour, intention—Fur, actuality—Metal) correspond to the modes of
declaration. Given his success in independently formulating propositional and first-order
logic in ways that are only relatively recently being recognized, perhaps it would not be such
a surprise were Peirce on the right track with his gamma system.

On that note we end the discussion of what a general modal HoTT should look like, and
turn now to put one particular version to use in a formulation of modern geometry.

# 5

## Spatial Types

## 5.1 Introduction

Leafing through Robert Torretti's book, *Philosophy of Geometry from Riemann to Poincaré* (1978), it is natural to wonder why, at least in the anglophone community, we have so little activity meriting this name today. Broadly speaking, we can say that any philosophical interest in geometry shown here is directed at the appearance of geometric constructions in physics, without any thought being given to the conceptual development of the subject within mathematics itself. This is in part a result of a conception we owe to the Vienna Circle and their Berlin colleagues that one should sharply distinguish between *mathematical* geometry and *physical* geometry. Inspired by Einstein's relativity theory, this account, due to Schlick and Reichenbach, takes mathematical geometry to be the study of the logical consequences of certain Hilbertian axiomatizations. For its application in physics, in addition to a mathematical geometric theory, one needs laws of physics and then 'coordinating principles' which relate these laws to empirical observations. From this viewpoint, the mathematics itself fades from view as a more or less convenient choice of language in which to express a physical theory. No interest is taken in which axiomatic theories deserve the epithet 'geometric'.

However, in the 1920s this view of geometry did not go unchallenged, as Hermann Weyl, similarly inspired by relativity theory, but working as a mathematical physicist, was led to very different conclusions. His attempted unification of electromagnetism with relativity theory in 1918 was the product of a coherent geometric, physical and philosophical vision, the latter inspired by his familiarity with the works of Fichte and Husserl. While this unification was not directly successful, it did give rise to modern gauge field theory. Weyl, of course, also went on to make a considerable contribution to quantum theory. As for Einstein himself, although he gave initial support to Moritz Schlick's account of his theory, he later became an advocate of the thesis that mathematics provides important conceptual frameworks in which to do physics:

> Experience can of course guide us in our choice of serviceable mathematical concepts;
> it cannot possibly be the source from which they are derived; experience of course
> remains the sole criterion of the serviceability of a mathematical construction for
> physics, but the truly creative principle resides in mathematics. (Einstein 1934, p. 167)

*Modal Homotopy Type Theory: The Prospect of a New Logic for Philosophy*. David Corfield, Oxford University Press (2020). © David Corfield. DOI: 10.1093/oso/9780198853404.001.0001

We may imagine, then, that an important chapter in any sequel to Torretti's book would describe both Schlick and Reichenbach's *and* Weyl's views on geometry. This is done in Thomas Ryckman's excellent *The Reign of Relativity* (2005), where the author also discusses further overlooked German-language philosophical writings on geometry from the 1920s, this time by Ernst Cassirer. Cassirer's extraordinary ability to assimilate the findings of a wide range of disciplines sees him discuss the work of important mathematicians such as Klein, Steiner, Dedekind and Hilbert.

Ryckman ends his book with a call for philosophical inquiry into what sense a *geometrized physics* can have today, to emulate the above-mentioned work from the interwar period. And he is not alone in thinking that this was a golden age. Indeed, there is by now an impressive concentration on this era. Today, in Ryckman and similar-minded thinkers, such as Michael Friedman and Alan Richardson (see the contributions of all three in Domski and Dickson (2010)), we find fascinating discussions about these themes. However, these discussions by themselves are unlikely to give rise today to the kind of primary work that they are interpreting from the past. It is one thing to make a careful, detailed study of the interweaving of philosophy, mathematics and physics of a bygone period, quite another to begin to take the steps necessary for a revival of such activity in the present.

I introduced Michael Friedman in §5 of Chap. 2 as a philosopher who, in contrast to Quine, proposed a hierarchically layered view of scientific theory, where empirical laws cannot even be expressed meaningfully without the assumption of constitutive principles, which themselves may well have required advances in mathematics to be expressed. Think of Newton's Law of Gravitation, which makes empirical claims when understood within the framework of his laws of motion, which themselves were articulable through his mathematical advances in formulating the differential calculus.

Friedman (2001) points out two connected, yet somewhat distinct, activities dealing with mathematical physics which might be called 'philosophical'. One, termed 'metascientific', is much as Weyl does, reconceiving the idea of space and thereby generating foundational advances, allowing the opportunity for new physical theories to be expressed. Metascience is typically undertaken by philosophically informed scientists, such as Riemann, Helmholtz, Poincaré, and Einstein. By contrast, the other activity is much as Cassirer and the Vienna Circle did, reflecting on the broader questions of the place of mathematics and science in our body of knowledge in light of important events in the histories of those practices. While there spontaneously arises work of the first kind in any era, work of the second kind requires a philosophical orientation which may be lost. One very obvious difference is that today we have so few philosophers emulating Cassirer by keeping abreast of the mathematics of the recent past. This simply must change if we are to generate forms of discussion to parallel those of the 1920s. For too long, philosophy has thought to constrain its interest in any current mathematical research largely to set theory, when it has long been evident that it offers little or nothing as far as many core areas of mathematics are concerned, and especially the mathematics needed for physics. Casting the differential cohomology of modern quantum gauge theory in set-theoretic clothing would do no favours for anyone. So, with some notable exceptions, such as Marquis (2008) and McLarty (2008), we let the bulk of mainstream mathematical research pass us by.

However, there are reasons to be hopeful. I shall argue in this chapter that our best hope in reviving a 1920s-style philosophy of geometry lies in following what has been happening at the cutting edge of mathematical geometry over the past few decades, and that while this may appear a daunting prospect, we do now have ready to hand a means to catch up rapidly. These means are provided by what is known as *cohesive homotopy type theory*, a variety of the modal HoTT that we studied in Chap. 4.

As has been explained earlier in the book, plain HoTT provides the syntax for theories which can be interpreted within $(\infty, 1)$-toposes. The basic shapes of mathematics are now taken to be 'homotopy $n$-types'. However, these are not sufficient to do what needs to be done in modern geometry, and especially in the geometry necessary for modern physics, since we need to add further structures to express continuity, smoothness, and so on. As we add extra properties and structures to $(\infty, 1)$-toposes, characterized by qualifiers—local, $\infty$-connected, cohesive, differentially cohesive—increasing amounts of mathematical structure are made possible internally. The work of Urs Schreiber (2013) has shown that cohesive $(\infty, 1)$-toposes provide an excellent environment to approach Hilbert's sixth problem on axiomatizing physics, allowing the formulation of relativity theory and all quantum gauge theories, including the higher-dimensional ones occurring in string theory.

The concept of *cohesiveness* in Schreiber's sense arose from earlier formulations of it in terms of a pattern of monads and comonads on *ordinary* toposes by William Lawvere (2007), motivated in turn by philosophical reflection on geometry and physics. Schreiber's claim, however, is that for these concepts to take on their full power they must be extended to the context of *higher* topos theory, that is, the theory of $(\infty, 1)$-toposes, where differential cohomology finds its natural setting. Now, rather than the mathematics necessary for physics being viewed as elaborate and unprincipled, as it often is at present from a set-theoretic perspective, we can see the simplicity of the necessary constructions through the universal constructions of higher category theory. If this ambitious project pans out, for instance by allowing the identification and expression of M-theory, the elusive theory which holds the key to the web of dualities uniting versions of string theory (see Fiorenza, Sati and Schreiber 2019), this would mark an excellent example of Friedman's schema. Higher topos theory makes possible the formulation of differential cohomology, which in turn makes possible the expression of higher quantum gauge field theory. Modal HoTT can express constructions within higher topos theory, in particular geometric morphisms between pairs of higher toposes, and it owes its origins to a host of philosophical reflections.

In a single chapter it will only be possible to outline the kind of work that is necessary to fill in the spaces we have left ourselves. We have accumulated interpretative debts all the way through this book, and more will surely follow from the brief indications given here. At the same time, in that mathematics finds itself once again undergoing enormous transformations in its basic self-understanding, it is important as philosophers to take this opportunity to remind ourselves that we should provide an account of mathematical enquiry where such changes are to be expected. It is striking that, as we shall see, Hegel should be found informing both those wishing to characterize the dynamic growth of mathematical physics, and those striving to refashion the very concepts of modern geometry itself.

## 5.2  Current Geometry

If any reassurance is needed that geometry is alive and well today, we need only look at the variety of branches of mathematics bearing that name which are actively being explored:

> Algebraic, differential, metric, symplectic, contact, parabolic, convex, Diophantine, tropical, conformal, Riemannian, Kähler, Arakelov, analytic, rigid analytic, global analytic, and so on.

Continuing our search, we find noncommutative versions of some of the above, 'derived' versions, and so on. Indeed, there has never been so much 'geometric' research being carried out as there is today, from constructions that Gauss or Riemann might have recognized to ones which would seem quite foreign. So the question arises of whether this list provides just a motley of topics which happen to bear the same name, or whether there is something substantial that is common to all of them, or at least many of them.

Evidence for the latter option comes from people still making unqualified use of 'geometry' and its cognates to mean something. Some such uses are informal, as in the following example:

> The fundamental aims of geometric representation theory are to uncover the deeper geometric and categorical structures underlying the familiar objects of representation theory and harmonic analysis, and to apply the resulting insights to the resolution of classical problems. (MRSI 2014)

Other uses are technical, such as where Jacob Lurie (2009b) uses the term 'geometry' to name a certain kind of mathematical entity: here a small $(\infty, 1)$-category with certain additional data. The question then arises as to what features of these structures make Lurie single them out as geometries. At first glance, this seems a rather technical matter; let us return to it once we have some motivation from the past.

Something that would have seemed novel to Gauss and Riemann, and which might give rise to doubts concerning the unity of geometry, is the thorough injection of spatial ideas brought into algebraic geometry by Alexandre Grothendieck in the 1960s. By the late 1800s it was already known of the collection of complex polynomials, $\mathbb{C}[z]$, that the space these functions are defined on could be recaptured from the algebraic structure of the collection itself. $\mathbb{C}[z]$ forms a *ring*, a collection endowed with an addition and a multiplication, and it is possible to construct an associated space of points corresponding to its *maximal ideals*. These are the ideals generated by $(z - a)$ for each $a \in \mathbb{C}$, so the collection of those polynomials with $(z - a)$ as a factor. Picking up on the complex function field/algebraic number field analogy of Dedekind and Weber, as developed by André Weil in his Rosetta Stone account (see Corfield 2003, Chap. 4), it was then shown that even apparently non-functional rings, such as the ring of integers and others encountered in arithmetic, might be treated likewise. Grothendieck's scheme theory (see McLarty 2008) provided such a space, which in the case of the integers is denoted $Spec(\mathbb{Z})$, again constructed out of maximal ideals.[1] Now an

---

[1] In general, scheme theory takes all *prime* ideals as points, the maximal ideal corresponding to the *closed* points.

integer is considered as a function defined at each prime, a point in $Spec(\mathbb{Z})$, as the function $n(p) \equiv n(mod\, p)$. Where this differs from the complex function case is that while the values of 'integer-as-function' still land in a field, here the field varies according to the point where the function is evaluated: $\mathbb{F}_p$ as $p$ varies. This suggests a non-homogeneous space whose points are not identical.

This attempt to geometrize arithmetic is not an empty game. It feeds through to mathematical practice, as can readily be discovered thanks to the growth of online informal discussion.

> I like to picture $Spec\,\mathbb{Q}$ as something like a 2-manifold which has had all its points deleted. The extra complication is that what we think of as the points are actually very small circles. So it's really a three manifold with all of the loops inside it deleted.
>
> For example, let's look first at function fields. $Spec\,\mathbb{C}[z]$ is just the complex line $\mathbb{C}$. As we start inverting elements of $Spec\,\mathbb{C}[z]$, as we must do to make $Spec\,\mathbb{C}(z)$, the effect on the spectrum is to remove bigger and bigger finite sets of points. The limit is where we remove all the points and we're just left with some kind of mesh.
>
> If we had started with a Riemann surface of genus $g$, then we'd be left with a mesh of genus $g$, a surface sewn out of the cloth from which fly screens for windows are made. If we want to recover the original surface from the surface mesh, we just put it out back in the shed for a while and let the mesh fill up with dirt. This is just the familiar fact that a (smooth compact, say) Riemann surface can be recovered from the field of meromorphic functions on it.
>
> If we replace $\mathbb{C}$ by a finite field $\mathbb{F}$, then everything is the same but what we thought of as the point is now a very small circle, and so our original surface reveals itself to be a 3-manifold fibered over a very small circle when we zoom in. And when we delete points, we're really deleting not just single-valued sections of this fibration but also multivalued sections. So $Spec\,\mathbb{F}[z]$ is some kind of 3-manifold fibered over the circle with all the loops over the base circle deleted.
>
> For the passage from $\mathbb{Z}$ to $\mathbb{Q}$, I don't have anything better to say than that it's sort of the same but there's no base circle. We're just removing lots of loops from a 3-manifold. Maybe some should be seen as bigger than others, corresponding to the fact that there are prime numbers of different magnitudes. (Borger 2009)

Here we see Borger passing across each of the three columns of Weil's Rosetta Stone.

Now, not only do we find a geometrized arithmetic, but these ideas and constructions are the structural cousins of those appearing in cutting-edge physics, as we see in this comment by David Ben-Zvi:

> the geometric analog of a number field or function field in finite characteristic should not be a Riemann surface, but roughly a surface bundle over the circle. This explains the 'categorification' (need for a function-sheaf dictionary, which is the weak part of the analogy) that takes place in passing from classical to geometric Langlands—if you study the corresponding QFT on such three-manifolds, you get structures much closer to those of the classical Langlands correspondence. (Ben-Zvi 2014)

So we have *both* widespread current interest in classically geometric areas of mathematics and geometric approaches to other areas, including arithmetic, *and* ways of thinking about the subject matter expressed, at least informally, in a very visual language. Geometry as a whole is something larger than that which has application in mathematical physics, and applied mathematics more generally. Similar structures are now found to lie at the heart of number theory. But how to go about saying something satisfactorily general about geometry?

One might throw up one's hands at the task of bringing this wealth of subject matter under the umbrella of a straightforward description. To the extent that people try to do this, it is largely left to the doyens of mathematics. For example, Sir Michael Atiyah writes

> Broadly speaking, I want to suggest that geometry is that part of mathematics in which visual thought is dominant whereas algebra is that part in which sequential thought is dominant. (Atiyah 2003, p. 29)

Such a distinction is reminiscent of Kant, for whom space and time were considered to be the forms of sensibility associated with geometry and algebra, and yet Atiyah continues

> This dichotomy is perhaps better conveyed by the words 'insight' versus 'rigour' and both play an essential role in real mathematical problems. (Atiyah 2003, p. 29)

A more careful treatment is required here, since there seems nothing to object to in the idea of 'rigorous geometry' or 'algebraic insight'. We need to turn back the clock to when philosophical research directed itself towards the then current geometry.

## 5.3 Regaining the Philosophy of Geometry

What led to the demise of the philosophy of geometry in the English-speaking world? I think this can be attributed largely to the success of logical empiricism. Many of those dispersed from Germany and Austria in the 1930s were accepted into the universities of the USA, welcomed by existing empiricists such as Ernst Nagel. In a long paper published in 1939, Nagel uses the history of projective geometry to explain the new understanding of then modern axiomatic mathematics:

> It is a fair if somewhat crude summary of the history of geometry since 1800 to say that it has led from the view that geometry is the apodeictic science of space to the conception that geometry, in so far as it is part of natural science, is a system of 'conventions' or 'definitions' for ordering and measuring bodies. (Nagel 1939, p. 143)

> The distinction between a pure and an applied mathematics and logic has become essential for any adequate understanding of the procedures and conclusions of the natural sciences. (Nagel 1939, p. 217)

So axiom systems are proposals for stipulations. As pure mathematics, they are to be studied for their logical properties. Some of them may be found to be appropriate for the expression of scientific laws, which may then be used in applied sciences. This is made possible by *coordinating principles* which tie the scientific laws to empirical measurements. For example, Riemannian geometry allows for the expression of Einstein's field equations, which can be coordinated to observation by stipulating that light follows null geodesics.

While now the ideational content of mathematics is left to one side, other tasks fall to the philosopher of mathematics:

> the concepts of structure, isomorphism, and invariance, which have been fashioned out of the materials to which the principle of duality is relevant, dominate research in mathematics, logic, and the sciences of nature. (Nagel 1939, p. 217)

Had philosophers at least heeded this, more attention might have been paid to category theory, the language *par excellence* of structure, isomorphism and invariance, which emerged shortly after Nagel's paper, and which allows a deep understanding of the geometric duality he treated there as well as all other mathematical dualities (Corfield 2017d). As it was, a uniform treatment of mathematics as the logical consequences of definitions, or of set-theoretic axioms, came to prevail.

In the process, as Heis (2011) argues, two lines of thought from earlier in the century were being ruled out, each responding to other nineteenth-century developments in geometry:

1. What could be saved of Kantian philosophy given the appearance of non-Euclidean geometry, and then Riemannian geometry? What are the conditions for spatial experience?
2. How should we understand the ever-changing field of geometry given the introduction of ideal elements, imaginary points, and so on?

Let us take each of these in turn.

### 5.3.1  Weyl: The Essence of Space

With an ever-expanding variety of geometries emerging through the nineteenth century, it became implausible to maintain with Kant that our knowledge of Euclidean geometry is a priori. Helmholtz had argued that because empirical measurement requires that objects undergo only 'rigid motions', we can work out which geometries are presupposed by our physics. He concluded that only those spaces which possessed the property of constant curvature were permissible. With the contribution of the technical expertise of Sophus Lie, this line of research resulted in the Helmholtz–Lie theorem, characterising Euclidean, elliptic and hyperbolic geometries.

Research such as this was certainly discussed by philosophers. Indeed, Russell, and later Schlick and Reichenbach, responded to Helmholtz's work, but perhaps the most profound response came from Hermann Weyl. After the success of Einstein's general theory of relativity, Helmholtz's results were evidently far too limited, pseudo-Riemannian manifolds of varying curvature providing the spaces for the theory. Weyl, inspired by Husserl and perhaps more profoundly by Fichte (see Scholz 2005), sought to discern 'the essence of space'. In a letter to Husserl, he wrote

> Recently, I have occupied myself with grasping the essence of space [das Wesen des Raumes] upon the ultimate grounds susceptible to mathematical analysis. The problem accordingly concerns a similar group theoretical investigation, as carried out by Helmholtz in his time. (Ryckman 2005, p. 113)

Weyl conceived of spaces in which it was only possible to compare the lengths of two rods if they were situated at the same point.

> Only the spatio-temporally coinciding and the immediate spatial-temporal neighborhood have a directly clear meaning exhibited in intuition. (Weyl 'Geometrie und Physik', quoted in Ryckman 2005, p. 148)

Along the lines of Helmholtz and Lie, this led him to prove a group-theoretic result: The only groups satisfying certain desiderata (involving the 'widest conceivable range of possible congruence transfers' and a demand for a single affine connection), are the special orthogonal groups of any signature with similarities, $G \simeq SO(p, q) \times \mathbb{R}^+$ (Scholz 2011).

There is an interesting story to be told here of how Einstein and other physicists found implausible the possibility allowed by this geometry that rods of identical lengths as measured at one point, if transported along different paths to a distant point, might have different resulting lengths. One usually tells the story of how the beauty of the mathematics got the better of Weyl, and how physicists eventually uncovered what was good about the idea while modifying his original idea to allow a $U(1)$ gauge group. This story needs to be told in a much more nuanced way (see Giovanelli 2013), and in any case is complicated by the survival of Weyl's original idea in forms of conformal gauge theory.

In any case, Weyl himself later became sceptical of this kind of mathematical speculation about the geometry required for physics that had so consumed him in his earlier years. With the demise of other philosophical attempts to study our a priori geometric intuition, for example, Carnap's doctoral thesis on how our intuitive concept of space was required to be $n$-dimensional topological space, such attempts largely came to an end. We will take a look at more recent 'metascientific' kinds of work, but first let us turn to Friedman's other form of philosophical research.

### 5.3.2 Cassirer: Beyond Intuition

In an unusual paper that we briefly considered in §4.2.1, which was translated into English and published the year before his death, Ernst Cassirer (1944) argued for an important connection to be seen between Felix Klein's Erlangen Program and our everyday perception. Whereas Klein had given a presentation of many forms of geometry as the study of invariants of space under the action of groups of symmetry, Cassirer saw the seeds for this idea in our abilities to perceive the invariant size, colour and shape of objects under varying viewing conditions. Now, evidently these abilities are rooted in our distant evolutionary past, and yet the full-blown mathematical idea had only crystallized in modern mathematical thinking around 1870.

Klein's ideas on geometry marked an important stage in the course of a revolutionary century for geometric thought. Not only had the range of geometries been extended from the single Euclidean geometry to hyperbolic and elliptical forms, but there had been many kinds of extension of the notion of space by the introduction of elements that seemed to lead us away from the intuitively familiar. For example, the non-degenerate conic sections had been unified as curves of degree 2 in complex projective space, brought about by the addition of 'points at infinity' and complex coordinates. Now all circles were seen to pass through two imaginary points at infinity, and so 'intersect' there.

Such forays beyond the intuitive led those wishing to retain what they took to be valuable in Kant to take a different tack. As Heis (2011) convincingly shows, the neo-Kantian Cassirer had to come to terms with just such developments. This is evident in his later work:

> It is hence obvious that mathematical theories have developed in spite of the limits within which a certain psychological theory of the concept tried to confine them. Mathematical theory ascended higher and higher in order to look farther and farther. Again and again it ventured the Icarian flight which carried it into the realm of mere 'abstraction' beyond whatever may be given and represented in intuition. (Cassirer 1944, p. 24)

But then without any firm rootedness in intuition, what provides us with guidance that our 'Icarian flights' are heading in the best direction? This issue is addressed by Cassirer throughout his career, and is answered by him in terms of the unity of the history of the discipline.

> Though a properly Neo-Kantian philosophy of mathematics will appreciate that mathematics itself has undergone fundamental conceptual changes throughout its history, such a philosophy will also have to substantiate the claim that the various stages in the historical development of mathematics constitute *one history* ... we can say that they [mathematicians] were studying the same objects only because we can say that they are parts of the same history. (Heis 2011, p. 768)

It is worth quoting Cassirer at length on this point:

> it is not enough that the new elements should prove equally justfied with the old, in the sense that the two can enter into a connection that is free from contradiction—it is not enough that the new should take their place beside the old and assert themselves in juxtaposition. This merely formal combinability would not in itself provide a guarantee for a true inner conjunction, for a *homogeneous logical structure of mathematics*. Such a structure is secured only if we show that the new elements are not simply adjoined to the old ones as elements of a different kind and origin, but the new are *a systematically necessary unfolding of the old*. And this requires that we demonstrate a primary logical kinship between the two. Then the new elements will bring nothing to the old, other

than *what was implicit in their original meaning*. If this is so, we may expect that the new elements, instead of fundamentally changing this meaning and *replacing* it, will first bring it to *its full development* and *clarification*. (Cassirer 1957, p. 392)

If one can hear an overtone of Hegelian thought here, this is not surprising. In the introduction to his third volume of *The Philosophy of Symbolic Forms*, Cassirer explained the debt to Hegel as shown by the subtitle of the book—*The Phenomenology of Knowledge*:

> The truth is the whole—yet this whole cannot be presented all at once but must be unfolded progressively by thought in its own autonomous movement and rhythm. It is this unfolding which constitutes the being and essence of science. The element of thought, in which science is and lives, is consequently fulfilled and made intelligible only through the movement of its becoming. (Cassirer 1957, p. xiv)

This line of thought sits very happily with the idea that important developments in a discipline allow its history to be written so as to make best sense of what was only obscurely seen in the past, and how perceived obstacles or limitations were overcome; in other words, a history of rational unfolding out of an older stage. As I argue in Corfield (2012), we find this position very well expressed by the moral philosopher Alasdair MacIntyre. A similar idea is expressed by Friedman's 'retrospective' rationality (2001).

In the anglophone revival (Friedman, Ryckman, Richardson, Heis and Everett) of interest in Cassirer of recent years there has been particular focus on the place of the 'constitutive' and the 'regulative' in his account of the progress of science. This amounts to rival interpretations of the relative importance for Cassirer of the prospective overcoming of limitations within a discipline and the retrospective rationalization of its course, eventually as seen from an ideal future point. However these debates turn out, it is intriguing, then, to see what we might call a further strand added in the 1944 paper, that in mathematics we may devise concepts which owe their origin to unnoticed cognitive structures. In the case of the Erlangen Program

> the mathematical concepts are only the full actualisation of an achievement that, in a rudimentary form, appears also in perception. Perception too involves a certain invariance and depends upon it for its inner constitution. (Cassirer 1944, p. 17)

Taken together, we can see in the work of Weyl and Cassirer just how far we are here in attitude towards mathematical geometry from what was bequeathed to us by the logical empiricists. Heis quotes Hans Reichenbach:

> It has become customary to reduce a controversy about the logical status of mathematics to a controversy about the logical status of the axioms. Nowadays one can hardly speak of a controversy any longer. The problem of the axioms of mathematics was solved by the discovery that they are definitions, that is, arbitrary stipulations which are neither true nor false, and that only the logical properties of a system—its consistency, independence, uniqueness, and completeness—can be subjects of critical investigation. (Heis 2011, p. 790)

He very aptly writes 'One could hardly find a point of view further from Cassirer's own' (Heis 2011, p. 790). Indeed so, but now to be true to Cassirer's spirit we should try to work out our own position on the rationality of mathematical enquiry in the process of coming to frame what has been happening in the mathematics of the recent past. That this has typically not been felt to be a requirement of philosophy makes this no easy task, but we should try to make a start anyway.

## 5.4 Capturing Modern Geometry

Even to summarize one particular line of development here will not be easy. Along with Grothendieck's invention of scheme theory, mentioned in §5.2, we would also need to talk far more about topos theory than we have to date. Grothendieck introduced the categories called *toposes* or *topoi* as part of his radical redevelopment of algebraic geometry through the 1960s. In part, this was to provide a better control over certain kinds of spaces and their subspaces. Having made the move to find spaces—the aforementioned *schemes*, underlying the rings of algebraic geometry, the natural next step was to provide them with *topologies*: ways to understand how their subspaces fit together with each other. But since the topologies available, such as the *Zariski* topology, were too coarse, Grothendieck adopted a different strategy. He realized that you can work with a space more effectively by understanding which other similar kinds of space *cover* it suitably. At the same time, he also saw that embedding a collection of spaces into a broader environment with nicer properties would provide greater traction. The extra objects necessary to provide these nicer properties might appear to be strange as spaces at first, but the benefits were substantial. This has become a principle of current mathematics: better to have a *nice* category of objects, than a category of *nice* objects. The resulting environments in the situations he was considering are what he termed *toposes*. As we have mentioned before, toposes have a great many pleasant properties, including the possession of all limits and colimits, providing ways to intersect, join, and so on.[2]

So an account of topos theory could take the form of a story of a natural unfolding. Indeed, we could report the originator's own words:

> one can say that the notion of a topos arose naturally from the perspective of sheaves in topology, and constitutes a substantial broadening of the notion of a topological space, encompassing many concepts that were once not seen as part of topological intuition ... As the term 'topos' itself is specifically intended to suggest, it seems reasonable and legitimate to the authors of this seminar to consider the aim of topology to be the study of topoi (not only topological spaces). (Grothendieck and Verdier 1972, p. 302)

Now important work through the middle years of the twentieth century had involved the construction of *homology* and *cohomology* theories, ways to assign consistently algebraic

---

[2] The duality we noted earlier between a geometric space and the algebra of functions defined upon it, as exemplified by schemes and rings, can be continued to toposes themselves and the sheaves that may be defined on them (see Anel and Joyal 2019).

entities to spaces, with a view to extracting useful information about these spaces. One target of Grothendieck was the Weil conjectures. This required the definition of a suitable *cohomology* in the setting of algebraic geometry, hence in the topos setting. The enormous body of his work through this period is contained in the publications known by their initials as EGA and SGA.

Much has been discovered about toposes since the 1960s. For one thing, topos theory became its own topic within category theory, and not merely of relevance to algebraic geometry. Close ties with the foundations of mathematics resulted from Lawvere and Tierney's axiomatization of *elementary* toposes in 1970. With this more general perspective, applications then followed in differential geometry, in an approach known as *synthetic differential geometry*.

The best-known reference work for topos theory is Peter Johnstone's *Sketches of an Elephant* (Johnstone 2002), a comprehensive work arranged in six parts, distributed through a planned three volumes. The title alludes to the story of six blind men feeling different parts of an elephant and thereby coming to very different conclusions about the subject of their investigation. The book exposes the various logical and geometric features of a topos. To date, only the first two volumes have appeared, back in 2002. The first section of the proposed third volume is titled *Homotopy and Cohomology*, which may strike some of its potential audience as out of date when it eventually appears, since it has become fairly common currency over recent years that the natural home for homotopy and cohomology is not in toposes, but rather in $(\infty, 1)$-toposes. As the reader who has made it this far through the book will guess, what is at stake is the question of identity. A topos, as something resembling the category of sets, has more limited resources to represent naturally the kinds of equivalence that geometric situations require. $(\infty, 1)$-toposes result from collecting together the $\infty$-groupoids mentioned in Chap. 3, with their higher equivalences.

We can gain a sense of what is stake by considering how Jacob Lurie motivates his 'Structured Spaces' paper (2009b) by way of an account of the passage to less restricted forms of Bézout's theorem. This is a result that goes back to the eighteenth century, involving just the kind of achievement of unity through addition of ideal elements that interested Cassirer. While it was known to Newton that the number of *real* solutions to the intersection of a pair of plane curves was bounded by the product of their degrees, by the nineteenth century we find a form of the result that states that two *complex* projective plane curves of respective degrees $m$ and $n$ which share no common component have precisely $m \cdot n$ points of intersection, counted with multiplicity. For instance, any two nonidentical conics meet four times, including at those two imaginary points at infinity, mentioned in section 5.3.2, through which all circles pass.

Lurie takes this result up, looking to understand it in terms of a *cohomological* operation to produce the intersection of two curves, by what is known as the cup product of the fundamental classes of the curves. This corresponds to the class of their intersection. But since this method does not work for non-transverse intersections, that is, intersections where the tangents of the curves involved coincide, some modifications are needed.[3]

---

[3] These involve first Grothendieck's construction of 'nonreduced' schemes, but then further, according to Lurie, an Euler characteristic involving the dimension of the local ring of the scheme-theoretic intersection plus various corrections.

Now, with these modifications in place, an interesting thing happens when we attempt to retain the fundamental result $[C] \cup [C'] = [C \cap C']$ in the very general setting where there may even be coinciding components. Here we need *derived* algebraic geometry.

> To obtain the theory we are looking for, we need a notion of generalized ring which remembers not only whether or not $x$ is equal to 0, but how many different ways $x$ is equal to 0. One way to obtain such a formalism is by categorifying the notion of a commutative ring. That is, in place of ordinary commutative rings, we consider categories equipped with addition and multiplication operations (which are encoded by functors, rather than ordinary functions). (Lurie 2009b, p. 3)

Lurie is drawing attention here to the passage from the *proposition* 'Is $x$ equal to 0?' to the *set* of ways in which it is equal. To do so is to take the first step up the infinitely tall hierarchy of $n$-types that we treated in 3.1.2. Lurie, in particular, was instrumental in this change in showing that most constructions of ordinary category theory have their analogues in the $(\infty, 1)$ setting, where instead of hom-sets between objects, we have hom-$\infty$-groupoids, or homotopy $n$-types.

Now, since when dealing with a pair of coinciding lines we need to make identifications in the form of isomorphisms, we find that

> These isomorphisms are (in general) distinct from one another, so that the categorical ring $C$ 'knows' how many times $x$ and $y$ have been identified. (Lurie 2009b, p. 4)

Of course, we never stop with a single step up this ladder, and eventually we seek further generalized forms of ring. Mathematicians call such things '$E_\infty$-ring spectra' and 'simplicial commutative rings'. The key lesson here is that to retain a simple formulation, we have to change our framework, for one thing here to allow *homotopic* weakening, equivalence rather than equality. In Cassirerian terms, this is forced upon us by the natural unfolding of the discipline.

One stage on this path to current geometry sees the study of homological algebra taking place within *derived* categories. In a work with Gaitsgory (Gaitsgory and Lurie 2019), Lurie explains the need to move beyond these in a section titled 'Motivation: Deficiencies of the Derived Category'. Thus, the derived category is 'not very well-behaved from a categorical point of view' (p. 66), since it generally lacks simple limits or colimits. Its deficiencies 'stem from the fact that we are identifying chain-homotopic morphisms ... without remembering how they are chain-homotopic' (p. 67). They add,

> It is possible to correct many of the deficiencies of the derived category by keeping track of more information. To do so, it is useful to work with mathematical structures which are a bit more elaborate than categories, where the primitive notions include not only 'object' and 'morphism' but also a notion of 'homotopy between morphisms.' (Ibid., p. 67)

Naturally, Lurie is not a lone voice in calling for this change of outlook. Bertrand Toën likewise gives an account of *derived algebraic geometry*:

> Derived algebraic geometry is an extension of algebraic geometry whose main purpose is to propose a setting to treat geometrically special situations (typically bad inter-sections, quotients by bad actions, ...), as opposed to generic situations (transversal intersections, quotients by free and proper actions, ...). (Toën 2014, p. 1)

He explains the need for 'homotopical perturbation' in Kuhnian terms,

> the expression *homotopical mathematics* reflects a shift of paradigm in which the relation of equality relation is weakened to that of homotopy. (Toën 2014, p. 3)

At the same time, he points the reader to the HoTT program as the new foundational language for this homotopical mathematics.

Now, despite this shift to what appears to be the more complex *derived* setting, familiar features are retained:

> Just as an ordinary scheme is defined to be 'something which looks locally like *SpecA* where *A* is a commutative ring', a derived scheme can be described as 'something which looks locally like *SpecA* where *A* is a simplicial commutative ring'. (Lurie 2009b, p. 5)

So, an apparently complicated space is being stuck together from pieces. This theme is taken up by Carchedi in a recent paper:

> we will make precise what it means to glue structured ∞-topoi along local homeomor-phisms (i.e. étale maps) starting from a collection of local models. This parallels the way one builds manifolds out of Euclidean spaces, or schemes out of affine schemes. Since we are allowing our 'spaces' to be ∞-topoi however, in these two instances we get much richer theories than just the theory of smooth manifolds, or the theory of schemes, but rather get a theory of higher generalized orbifolds and a theory of higher Deligne-Mumford stacks respectively. This same framework extends to the setting of derived and spectral geometry as well. (Carchedi 2013, p. 43)

It is not just algebraic geometry that demands this richer notion of space. So does theoretical physics. The moduli spaces of today's gauge field theories are often *stacks*, such as the moduli stack of flat connections for some gauge group. Higher gauge theory requires spaces with similar homotopic weakening to what we have seen above (Schreiber 2013).

The obvious point to be made is that all of this is just simply unthinkable without category theory. No category theory, and indeed no $(\infty, 1)$-category theory, and there is no modern geometry of this kind. On the other hand, it may strike the reader as rather daunting that we may need to get a good handle on what Lurie, Toën and Carchedi are doing with $(\infty, 1)$-toposes if they wish to make sense of the current situation in geometry.

If we recall Friedman's scheme of metascientific work leading up to a revolution followed by philosophical interpretative work to make sense of it, we might say that the cycle was largely broken through the twentieth century. Even the lessons of the seventy-year-old category theory are still very far from having been absorbed within philosophy. There have been many contributions made over the decades, but not the kind of sustained work that would make it matter of course for someone entering into a career in philosophy of mathematics to know the basics of category theory. It is hard to imagine that any other than very specialist philosophers will come to terms with the rise of a current geometry if this requires the absorption of the details of the path from Grothendieck to Lurie. We shall now see whether, with the resources of HoTT to hand, and in particular with the extension of it to a variety of modal HoTT, we can to some degree *leap-frog* this period.

## 5.5 Geometry in Modal HoTT

A common criticism of set theory as a foundational language has been its *unnatural* treatment of certain concepts, none more so than spatial ones. Where the set-theoretic approach can only look to build up a set that behaves like a space from the dust of its points, homotopy type theory can at least build into its basic entities something path-like in the collection of identities between terms. So as the latest geometry turns to the *n*-types belonging to $(\infty, 1)$-toposes, an approach to them which works *analytically* by defining such entities in terms of set theory will necessarily be dealing with very complicated formulations.

Thus we might hope for a *synthetical* formulation of modern geometry in HoTT, since its types are implicitly defined to behave as *n*-types, with all of their internal identity structures catered for. HoTT sees a set as a 0-type, something requiring additional characterization as a truncated form of *n*-type, whereas set theory sees an *n*-type as an elaborate set with a great deal of structure. This is certainly an advantage for HoTT, but we must be careful here since

> although it is common in homotopy type theory to use terminology borrowed from topology such as 'path' and 'circle', these words have *a priori* nothing to do with their topological versions *which can also be defined inside of type theory* . . . This frequently causes confusion among newcomers to homotopy type theory, who struggle to understand the meaning of 'path' because it both is, and is not, like the topological concept after which it is named. (Shulman 2018a, pp. 3–4)

Indeed, homotopy type theory is a synthetic theory of *structure* or of *higher equality*, not of *spaces*. Something more than HoTT is needed to bring the spatial properly within its grasp. This point can be seen clearly from the two versions of circle available in the theory. As we saw in Chap. 3, the native one uses *higher inductive types* to define a type, *S*, with a designated element known as its base point, *base*, along with a designated identity element, *loop*, in $Id_S(base, base)$. It is natural to think of this as a point with a loop which may be iterated indefinitely, just as one might wind a string around a circular reel. This circle type can then be used to do a synthetic form of algebraic topology, such as finding the type of maps from *S* to itself as an intrinsic form of the fundamental group of the circle, which is then calculated to be isomorphic to the integers (Licata and Shulman 2013).

The second approach is to define a circle as a locus of points in the plane. This is a much more elaborate, and less intrinsic, construction in the type theory, which first requires the construction of the reals, before forming the plane and then the subtype of this which is the circle, in much the way one would ordinarily produce a circle in analytic geometry. In the bare type theory there is no connection between these two 'circles'. We therefore need to add features that will represent how points are continuously connected, but we look to do this in *synthetic* fashion. This is what *cohesive* HoTT sets out to achieve, rather than imposing continuity conditions on the discrete. Once we have introduced its modalities, we will have the means to relate our two circles.

The kind of geometric modalities we will need arise from a string of *four* adjunctions, which in turn will generate an adjoint *triple* of monads and comonads. Up until now we have only seen adjoint pairs in which the monad appears as left adjoint and the comonad as right, $\bigcirc \dashv \square$. We will need to look also at the opposite situation, $\square \dashv \bigcirc$. We can begin by returning to the possibility/necessity construction of Chap. 4,

$$(\sum_f \dashv f^* \dashv \prod_f) : \mathbf{H}/X \overset{\overset{f_!}{\rightarrow}}{\underset{f_*}{\rightarrow}} \mathbf{H}/Y,$$

and noting again, as we did in §4.4, that compositions may be made in a different order, generating a monad/comonad pair on $\mathbf{H}/Y$. In computer science, in the case of terminal maps, $f : W \rightarrow \mathbf{1}$, these go by the name:

*writer comonad $\dashv$ reader monad,*

where we now have the comonad as *left* adjoint.

As we saw, the *reader monad* has the effect of sending a type, $A$, to the type $[W, A]$. A calculation in this type will result in a function which can *read* the state of the environment, $w : W$, to generate a value in $A$. The dual comonad is known as the *writer comonad*, sending a type, $B$, to the type $W \times B$. A calculation resulting in this type not only provides a value of $B$, but also *writes* a value of $W$. If $W$ happens to be a monoid, these *effects* can be aggregated to keep a 'running total' of the effects of computation.

The reader monad and writer comonad are not *idempotent*, in the sense that applying them twice is identical to applying them once, but they provide a first glimpse into the second orientation we need. The way we read the adjoint triple in 4.2.1 for the terminal map $W \rightarrow \mathbf{1}$, we had the domain of invariant or constant types embedding in a single way into the world-dependent types, and then two projections from the latter to the constant types, corresponding to what happens at *some* world and what happens at *all* worlds. For the other orientation, now we're thinking about embedding certain world-dependent types in two different ways into invariant types. We can make this clearer with a pair of more basic examples provided by William Lawvere (2000).

(i) *Adjoint modalities where the monad is the left adjoint, $\bigcirc \dashv \square$:*
Consider the simple case of the integers included in the real numbers, $I : \mathbb{Z} \hookrightarrow \mathbb{R}$, both regarded as linear orders. As categories, there is an arrow from one number to another, that is, between objects, precisely when the first is less than or equal to the second. This inclusion has as left adjoint the ceiling function, *ceil* $: r \mapsto \lceil r \rceil$, which

rounds up a real number to the nearest integer. Right adjoint is the floor function, *floor* : $r \mapsto \lfloor r \rfloor$, which rounds down to the nearest integer. Then $i \cdot floor \dashv i \cdot ceil$ are adjoint modalities.

Here, as with the case of necessity and possibility, we have one injection and two projections. There is only one *moment*, that of being integral, and general elements, here real numbers, project to elements *pure* according to this moment, here integers, in two different ways.

(ii) *Adjoint modality where the monad is the right adjoint,* $\Box \dashv \bigcirc$:
Consider the integers, $\mathbb{Z}$, as a partially ordered set and then the two maps, $\mathbb{Z} \to \mathbb{Z}$, the first of which picks out the even integers, *even* : $n \mapsto 2n$, and the second of which picks out the odd integers, *odd* : $n \mapsto 2n + 1$. These maps form an adjoint triple with the map which sends any integer to the integer part of half of itself, $\lfloor -/2 \rfloor$ : $n \mapsto \lfloor n/2 \rfloor$. Then from this triple adjunction, *even* $\dashv \lfloor -/2 \rfloor \dashv$ *odd*, we can generate an adjoint modality, *Even* $\dashv$ *Odd*, where *Even* has the effect of sending an integer to the equal or next lower even integer, and *Odd* has the effect of sending an integer to the equal or next higher odd integer. For instance, $Even(5) = Even(4) = 4$, while $Odd(5) = Odd(4) = 5$.

We have here a single projection and two injections. Lawvere describes the situation in terms of the opposite pair of *moments* of evenness and oddness. There is an equivalence between elements which are *pure* according to each of the moments, so the even and odd numbers are equinumerous as the images of distinct injections.

In sum, we have the two schemas, which in the situations we need to consider have been assigned technical terms:

- Two projections, one injection – $\bigcirc \dashv \Box$: *bireflective subcategory.*
- One projection, two injections – $\Box \dashv \bigcirc$: *essential subtopos.*

As we might expect from its sharing that pattern, it is easier to see the first scenario in terms of possibility and necessity. For instance, in the arithmetic case above you see a needle on a dial pointing between 4 and 5. You might be prepared to hazard an estimate of 4.6, but you could say more confidently that it's certainly at least 4 and possibly as much as 5.

What happens now is that we find an adjoint *quadruple* of functors which then compose to give an adjoint triple of modalities: $\bigcirc \dashv \Box \dashv \bigcirc$. There will not be space to go into much detail here, but let us begin at the ordinary 1-category level with Lawvere's notion of *cohesion* (Lawvere 2007) expressed as a chain of adjunctions between a category of spaces and the category of sets. If we take the former to be topological spaces, then one basic mapping takes such a space and gives its underlying set of points. All the cohesive 'glue' has been removed. Now, there are two ways to generate a space from a set: one is to form the space with the discrete topology, where no point sticks to another; the other is to form the space with the codiscrete topology,[4] where the points are all glued together into a single

---

[4] Sometimes called the *indiscrete* topology.

blob so that no part is separable, in the sense that there are only constant continuous maps from a codiscrete space to the discrete space with two points. Finally, we need a second map from spaces to sets, one which 'reinforces' the glue by reducing each connected part to an element of a set, the connected components functor, $\pi_0$:

$$(\pi_0 \dashv Disc \dashv U \dashv coDisc) : Top \to Set.$$

These four functors form an adjoint chain, where any of the three compositions of two adjacent functors $(U \circ coDisc, U \circ Disc, \pi_0 \circ Disc)$ from the category of sets to itself is the identity, whereas, in the other direction, composing adjacent functors to produce endofunctors on $Top$ $(coDisc \circ U, Disc \circ U, Disc \circ \pi_0)$ yields two idempotent monads and one idempotent comonad.

One adjoint modality, $Disc \circ U \dashv coDisc \circ U$, takes the form of one projection with two injections, first into the spaces with no cohesion and second into those with total cohesion. Lawvere (1994, p. 6) explains this adjunction as resolving the tension between a pure set being composed, on the one hand, of identical elements with no distinguishing properties, and yet, on the other, of different elements, resulting in 'a contradiction in a productive sense'.

The other adjoint modality, $Disc \circ \pi_0 \vdash Disc \circ U$, with its monad as left adjoint, involves two projections and one injection. The opposition expressed here is between an attraction, pulling together any connected points, and a repulsion, dismissing any connectedness. Any space, $X$, is situated between these versions of itself in that there are continuous maps, $Disc \circ U(X) \to X \to Disc \circ \pi_0(X)$.

What Schreiber does is to find analogous modalities generated by an adjoint quadruple between an $(\infty, 1)$-topos, $\mathbf{H}$, and the base $(\infty, 1)$-topos of $\infty$-groupoids, $\infty Grpd$:

$$(\Pi \dashv Disc \dashv \Gamma \dashv coDisc) : \mathbf{H} \to \infty Grpd.$$

The three induced adjoint modalities he calls shape modality $\dashv$ flat modality $\dashv$ sharp modality, and denotes them $\int \dashv \flat \dashv \sharp$. With $\int$ we can now relate those two presentations of circles earlier. Indeed, applied to the circle as a subset of the plane, this modality yields the circle as a higher inductive type. In a sense, this cohesive $\mathbf{H}$ can be seen as collecting together spaces modelled on a 'thickened' point, as though a point in a space coheres to its neighbours. This allows us to trace continuous paths through a space.[5]

The modality $\sharp$ can help make sense of the difference between *intensive* and *extensive* quantities. *Intensive* quantities, such as density, are those, $X$, for which the unit, $X \to \sharp X$, is a *monomorphism*, which means that the value of a function on $Y$ with values in $X$ is determined by its value on the points of $Y$, since we have $Hom(Y, X) \hookrightarrow Hom(Y, \sharp X) \cong Hom(\flat Y, X)$. So, for instance, if $Y$ represents the volume of some substance, then the density at any point of $Y$ is a value in $X$. *Extensive* quantities, such as mass, on the other hand, are such that $\sharp X = 1$, which means that $X$-valued functions on $Y$ vanish on its points and only receive a value from an extended region of $Y$, since we have $Hom(Y, 1) = Hom(Y, \sharp X) \cong Hom(\flat Y, X)$. In physics

---

[5] Discoveries concerning cohesion continue to be made. Charles Rezk (2014) has observed an instantiation relating to 'global equivariance'.

these extensive quantities will often appear as *differential forms*, which need to be integrated to deliver a value. Intensive quantities often appear as limiting ratios of extensive quantities as their supporting regions are reduced to a point, as with density defined as the ratio of mass to volume.

Now a very similar pattern repeats itself in the form of a further string of four adjunctions, this time between **H** and another $(\infty, 1)$-topos, corresponding to extending the thickened point infinitesimally. The three resulting adjoint modalities now comprise two comonads and one monad: $\square \dashv \bigcirc \dashv \square$. The existence of these two related sets of three adjoint modalities, termed *differentially cohesive* or *elastic*, is extraordinarily powerful, allowing the expression of a rich internal higher geometry, including Galois theory, Lie theory, differential cohomology and Chern–Weil theory, and allows for the synthetic development of higher gauge theory (Schreiber 2013). Already these modalities have been put to work in establishing formalized results in differential geometry using the proof assistant *Agda* (Wellen 2018). Finally, it is possible to add yet a further adjoint triple of modalities to the two mentioned, allowing for the expression of supergeometry, the setting for much current mathematical physics, where classical manifolds are replaced by those admitting additional Grassmann coordinates. Modalities here are described as *solid*, since they underpin the mathematics needed to describe the behaviour of electrons and other fermions, which governed by the Pauli exclusion principle gives rise to the solidity of matter in our world.

There remains plenty more interpretative work to be done in making these ideas more accessible, but for our purposes here let us just retain the monad of the second adjoint modality triple, denoted $\mathfrak{I}$ or sometimes $\int_{inf}$. As hinted at in Chap. 4, we can use $\mathfrak{I}$ and a space, $X$, much in the same way as we used $W \to 1$ to formulate a conception of possible worlds. Base change along the unit of the monad, $X \to \mathfrak{I}(X)$, and its right adjoint gives rise to a general concept of partial differential equations. In very much the same way as necessity was concerned with how to find counterparts in neighbouring situations/worlds, a differential equation can be thought to dictate how initial conditions are to be extended over infinitesimal neighbourhoods to form a solution. In fact, these situations are strongly analogous. The terminal map from an object of an $(\infty, 1)$-topos, **H**, can also be considered as the unit of a monad. This time the monad is the functor which sends every object to **1**. Its opposite is the comonad which sends every object to **0**, the initial object. Using these objects in bold face also as the names of the corresponding monad and comonad, we find that they form an adjunction, $\mathbf{0} \vdash \mathbf{1}$, so that for any object, $X$, there are morphisms, $\mathbf{0} \to X \to \mathbf{1}$. An intriguing story that Schreiber tells, building on ideas of Lawvere, is that this tower of adjoint triples can be explained as growing out of the opposition, $\mathbf{0} \dashv \mathbf{1}$. Just briefly, for the first step we can see $\sharp$ as a minimal reconciler of this initial opposition, in the sense that $\sharp\mathbf{0} \simeq \mathbf{0}$, it being necessary that $\sharp$ as a right adjoint would preserve the terminal object, **1**. The tower of modalities now proceeds via a continuing series of extensions.[6]

[6] It seems that in their treatment of the completeness of certain models for first-order modal logic, Awodey and Kishida are relying on a modality generated by a map analogous to the counit deriving from one of the modalities here, $\flat X \to X$, but for an ordinary topos. As they explain (Awodey and Kishida 2008, p. 161), their necessity comonad on the topos *Sets*/$|X|$ of sets indexed over a set $|X|$ is induced by the (continuous) map of spaces $id : |X| \to X$, from $|X|$ the discrete space with the points of $X$. This discrete space $|X|$ is just our $\flat X$.

ℑ turns out to be an important modality for us to continue the story from Lurie and Carchedi. So now, despite the apparently intimidating complexity of modern geometry, it is possible to maintain as Schreiber does that there remains a simplicity:

> It would seem to me that the old intuition, seemingly falling out of use as the theory becomes more sophisticated, re-emerges strengthened within higher topos theory ... Notably all those 'generalized schemes', 'étale infinity-groupoids' and so forth are nothing but the implementation of the old intuition of 'big spaces glued from small model spaces' implemented in homotopy theory ... I think it's a general pattern, in the wake of homotopy type theory we find that much of what looks super-sophisticated in modern mathematics is pretty close to the naive idea, but implemented internally in an ∞-topos. (Schreiber 2014b)

With homotopy type theory augmented by the addition of the first two adjoint triples, and in particular the infinitesimal shape modality, it is possible to describe *synthetically* what it is to be a 'formally étale morphism'. Such a mapping is to be used to display how a collection of *model spaces* is to be glued together smoothly in the right kind of way to form a given space. So now choosing types, $\{U_i\}$, as model spaces, then a general geometric space is a type, $X$, equipped with a map from a disjoint sum of model spaces of the form

$$\coprod_j U_j \longrightarrow X,$$

such that this map is *onto*, that is, it covers the whole of $X$, and it does so smoothly.[7] Think, for instance, of a number of flat patches being joined together with overlaps to cover all of a sphere.

We have arrived thus at a synthetic formulation of one of the very basic ideas of geometry. However intricate the 'spaces' of current mathematics, we never leave behind the essential idea of constructing them by gluing together specified model spaces. A patchwork of spaces is perhaps a feature of our everyday understanding of our surrounding space. It seems that when we work out a way to walk across a city, we have something like local maps in our mind which we can look to adjoin to provide us with a feasible route. With the addition of the modalities, homotopy type theory can express this gluing process straightforwardly, as it can the geometric constructions needed to formulate gauge field theories. Of course, there are many such basic ideas to be considered.

## 5.6  Simplicity and Representability in Modal HoTT

Over the sections 5.4 and 5.5 I have sketched some ideas from an extraordinarily ambitious body of mathematical, scientific and *metascientific* work. It may appear that by proposing

---

[7] In the case of schemes, one needs to modify slightly to *pro-étale* morphisms, in some sense a reflection of the less homogeneous nature of the spaces. Two very recent approaches to the pro-étale world go by the names *condensed* set and *pyknotic* set.

that we understand cutting-edge geometry, I risk being caught up with the changeable fashions of research, but let us not forget that these projects are rooted in the ideas of Grothendieck from many decades ago, and that later developments were foreseen to some considerable extent by him (see, for example, Grothendieck 1983). Current ideas thus emerge out of a vast body of prior work. Indeed, Toën motivates a section where he constructs 'a brief, and thus incomplete, history of the mathematical ideas that have led to the modern developments of derived algebraic geometry' as follows:

> As we will see the subject has been influenced by ideas from various origins, such as intersection theory in algebraic geometry, deformation theory, abstract homotopy theory, moduli and stacks theory, stable homotopy theory, and so on. Derived algebraic geometry incorporates all these origins, and therefore possesses different facets and can be comprehended from different angles. We think that knowledge of some of the key ideas that we describe below can help to understand the subject from a philosophical as well as from technical point of view. (Toën 2014, p. 6)

If some details will inevitably change, that $(\infty, 1)$-categories, and in particular $(\infty, 1)$-toposes, lie at the heart of modern geometry will very likely not. Indeed, many mathematicians are coming to see that they have freed themselves from earlier stifling ways of thinking. An enthusiastic expression of such sentiments appears in the thesis of Aaron Mazel-Gee:

> The many fussy details that arise when one attempts to use point-set techniques to work homotopy-coherently simply melt away: they were in fact irrelevant all along to the true and underlying mathematics, and their disappearance into the ambient machinery brings with it a harmony that is only possible when intuition and language are once again aligned. Thus, paradoxically, by *discarding* such emotional crutches as underlying sets and strict composition and by *embracing* the apparent chaos and uncontrol of homotopy-coherence, we acquire a measure of power of which previous generations of mathematicians could barely have dreamed. (Mazel-Gee 2016, p. 23)

Experience shows that it is no easy matter to dampen the suspicions of those who respond towards proposed new languages along the lines of *What can one say in the new language that cannot be said in the old one?* By what appear to the innovators to be forced constructions, the conservatives can relieve themselves of the burden of change. Remember that Newton himself avoided the use of his own calculus in the *Principia* in favour of something more closely approximating Euclidean geometry. His motivation seems to have been to ensure the appearance of continuity with the 'giants' on whose shoulders he stood, so as to repudiate the call for a radical break with the Ancients, so commonly expressed by the anti-scholastic voices of the seventeenth century. Of course, such a geometric approach didn't prevail for very long, and the benefits of the calculus, with Leibniz's notational variant, shone through over the course of time.

Scepticism in our story can occur at any point along the chain, so we might hear doubts from mathematicians that homotopical approaches to geometry are worthwhile or from physicists that they need higher topos theory to formulate cohomological accounts of string theory.[8] But let us now consider doubts from higher topos theorists that they need a formal language such as HoTT. Since modal HoTT is so young, it has not yet been sufficiently noticed to have attracted the sceptical glances of modern geometers.

It is clear that what is to count as a range of sufficiently fruitful developments springing from a new language to warrant adoption is something highly contested. Something novel, however, about the case in hand, the introduction of HoTT, is that we have a *pragmatic* rationale for its adoption. Voevodsky's main motivation was to develop a proof assistant to cope with the day-to-day reasoning of mathematicians working in homotopical mathematics. He claimed to have realized quickly that any system based on set theory could not possibly be made to work—there being too many encodings of homotopy equivalences to maintain. Hence his development of HoTT within the computer languages of Coq and Agda with its capacity to track homotopical equivalence already built in.

This has given rise to a body of work on *synthetic* homotopy theory. While a good number of constructions have been encoded in HoTT, and proofs of results derived in Coq and Agda, it may appear at first sight that, compared to the weighty tomes of classical theorists, these are humble offerings. It should be remembered, though, that any such result applies in a host of different settings, that is, in any higher topos. Varying the choice of the latter generates what the classical homotopy theorist would see as a range of related yet different theorems. Furthermore, what is a natural construction to a user of HoTT may not have directly occurred to a classical homotopy theorist. The following interesting case of this phenomenon is instructive.

A standard result in homotopy theory is the Blakers–Massey theorem, which concerns the connectedness of a map arising from an amalgamation of two other maps. The degree of connectedness of a space or map is complementary to the degree of truncatedness, the latter as measured by the level in the $n$-type hierarchy of Chap. 3. This theorem became a target for the synthetic homotopy theorists. First, a HoTT proof of Blakers-Massey was sketched and then a formalization in Agda constructed. This work caught the attention of a classical homotopy theorist, Charles Rezk, who was seeking a proof of the theorem as applying to any higher topos. Rezk translated the HoTT proof into ordinary mathematics, which now appears in his article 'Proof of the Blakers–Massey Theorem' (2015). Rezk remarks that his proofs

> are meant to be reverse engineered versions of proofs in homotopy type theory due to Lumsdaine, Finster, and Licata. The proof of Blakers-Massey given here is based on a formalization given by Favonia. (2015, p. 1)

The results alluded to were written up in Hou et al. (2016). The point of this story is that Rezk considered it worth the trouble to understand the HoTT proof, and indeed he declared it 'a most excellent proof!' (Schreiber et al. 2014). It is remarkable, then, that there was something to learn by a specialist homotopy theorist from a HoTT proof of a standard result constructed by a non-specialist. HoTT allows ideas to be encoded in all of their simplicity without the kinds of book-keeping measures that an analytic approach would require.

---

[8] Much more loudly these days, we hear from those who believe string theory itself is a wild goose chase.

Further developments of this HoTT proof, treated in Anel et al. (2017), exploit the generality of its interpretation in any higher topos. Furthermore, since truncations to $n$-types for a given $n$ are examples of *modalities*, and as observed above, truncatedness and connectedness are complementary, the authors were able to extend these HoTT proofs to a wide range of modalities. Particular cases of this extension were known in the literature, but, as they note, by following the HoTT construction 'the necessary input from classical homotopy theory has become almost invisible' (Anel et al. 2017, p. 3).

In light of what we discussed in §5.5, attention has now turned to the construction of computer-assisted proofs in *cohesive* homotopy type theory to handle geometric and topological results, rather than merely homotopic ones. Just as we have a philosophically satisfying calculus to handle properly equivalence as a refined form of equality, so we now have a calculus to handle cohesion. Three important references here are by Schreiber (2013), Shulman (2018a) and Wellen (2018). Wellen's thesis, from which his paper is drawn, tells us that all of the major theorems contained within were checked by the Agda system. As with set theory or any other formal system, we may expect there to be more human-friendly dialects in more-or-less close relationship to the formalism, from computer scientists' *sugaring* to mathematicians' *abuse of notation*. While a computer vouching for a construction is enormously reassuring, and a good way to root out slips in the informal argument, the conceptual benefits of a system will be gained to a large extent already in the informal version.

I am suggesting, then, that we see the simplicity of important homotopic and geometric ideas as expressed in HoTT and its variants, and the associated practical possibility of encoding results concerning them on a computer, as a sign that we are approaching the 'philosophical' essence of some of these ideas. Along with Michael Polanyi, we might hesitate before using the term *simplicity*:

> Hermann Weyl lets the cat out of the bag by saying: 'the required simplicity is not necessarily the obvious one but we must let nature train us to recognize the true inner simplicity.' In other words, simplicity in science can be made equivalent to rationality only if 'simplicity' is used in a special sense known solely by scientists. We understand the meaning of the term 'simple' only by recalling the meaning of the term 'rational' or 'reasonable' or 'such that we ought to assent to it', which the term 'simple' was supposed to replace. (Polanyi 1958, pp. 16–17)

But that it be possible for ideas to be expressible due to our new logic in a way that highlights more readily that they are *rational* or *reasonable* must count heavily in its favour.

## 5.7 Conclusion

I have sketched a broad canvas in this chapter. This is to some degree forced upon us by the state we are in where philosophy has drifted from its task. Had the course of philosophy after the famous Davos meeting (Friedman 2000) favoured Cassirer, we might have had a generation of philosophers keen to search for the emergence of new self-understandings in mathematics. Surely in that case category theory, and its higher forms, would have been absorbed much more fully into philosophical consciousness. With the emergence of HoTT, we may see this happen at last. What I have described in this chapter should suggest that there is a great deal of further work to be done in coming to understand *modal* extensions of HoTT, certainly the cohesive variety so far as geometry goes. It should also be noted that with a *linear logic* variant of homotopy type theory it is possible to express synthetically many aspects of the quantization of higher gauge theory (Schreiber 2014a).

We have seen Weyl-like metascientific work in the formulation of cohesive homotopy type theory, requiring a range of modalities to be added to the basic type theory. Unlike Weyl with Fichte, Schreiber follows Lawvere (1970, 1991) in finding inspiration in Hegel. As alluded to above, one can even tell a 'Hegelian' story starting from the adjoint opposition between **0** and **1**, rising through a process of 'Aufhebung' to the six modalities (Schreiber 2014a, §2.4), and beyond to the three modalities interpreted as capturing the supergeometry needed for dealing with fermions.

Contrast this with a different kind of use of Hegel by Cassirer and also Lakatos, a philosopher more familiar to the anglophone community. With the new framework for geometry in place, we should be able to tell the Cassirerian story of the unfolding of the past in mathematics and physics, as mathematicians such as Toën are inclined to do by themselves. Mathematics is to be understood by the fact that it constitutes a single tradition of intellectual enquiry. Ideas found at particular stages possess the seeds of later formulations, which retrospectively allow us to understand the findings of these earlier stages more profoundly.

Finally, as with Cassirer's observation about the seeds of the Erlangen Program lying within our perception, it is sometimes revealed during and after moments of synthesis in mathematics that there is a reliance on aspects of cognition, perception and language, which had possibly gone unnoticed. It would be worth exploring the detection of invariance under groups of transformations from the perspective of our type theory where the context is given by the type **B**G, as we discussed in §3.2.2. With the constructions of this chapter, we can look to invariance under actions by smooth groups, or *Lie* groups. Likewise, the idea of big spaces glued from small model spaces seems very basic. It is surely no accident that mathematicians speak of an 'atlas' to define a manifold, since an ordinary atlas provides a collection of maps which overlap. It seems likely we employ something like this in the cognitive maps by which we navigate our domain. Perhaps one of the true invariants of geometry has been found here.

# 6

. . ● . .

# Conclusion

*Modern logic, as I hope is now evident, has the effect of enlarging our abstract imagination, and providing an infinite number of possible hypotheses to be applied in the analysis of any complex fact. In this respect it is the exact opposite of the logic practised by the classical tradition.* (Bertrand Russell, *Logic As The Essence Of Philosophy*, 1914)

We have now reached the end of the tour through the construction of what I am proposing as philosophy's *new* new logic. Along the way, I hope to have given the reader sufficient glimpses of the ways in which philosophy can profit from taking to heart the language of modal homotopy (dependent) type theory: the discipline of a *type* theory, the flexibility of type *dependency*, the more refined *homotopic* notion of identity and a powerful range of *modalities*. In combination, the appearance of modal HoTT can provoke, in Russell's terms, the next great enlargement of our abstract imagination in philosophy. In the extract above, he gives us the impression of enjoying release from a straightjacket. The shift from subject–predicate syllogisms to a logic of relations did indeed allow far greater expressivity, notably through the extended use of quantifiers. However, the scope of useful application of his first-order logic in philosophy has proved to be too narrow and its guidance too weak. A logic for any domain should involve a mixture of freedom and constraint–freedom to explore productive possibilities, while also constraint to keep us from wandering down sterile *culs-de-sac* of our own making. One such lack of constraint in first-order logic can be seen in the way that by comparison with type theory it departs further in certain respects from syllogistic logic. Indeed, the general proposition '*All As are Bs*' which was rendered $\forall x(Ax \to Bx)$ by the tradition following Frege, is represented with less distortion by type theory as $f : \prod_{x:A} B(x)$, with no thought that the proposition has any relevance for the non-$A$. Maintaining type discipline is critical for the deployment of type theory, and a constraint well worth having.

A good example of helpful extended freedom comes with HoTT's liberation from the dominion of sets, so that types may display more intricate forms of identity. Taken in the *analytic* style, that is, defined in terms of sets, an $\infty$-groupoid is a complicated thing to build with all of its higher paths and coherence relations. Taken in the *synthetic* style, on the other hand, $\infty$-groupoids are defined implicitly via HoTT's refusal to admit a commonly applied restraint, namely, the axiom of the *uniqueness of identity proofs* (UIP), which stipulates that

*Modal Homotopy Type Theory: The Prospect of a New Logic for Philosophy*. David Corfield, Oxford University Press (2020). © David Corfield. DOI: 10.1093/oso/9780198853404.001.0001

any two proofs that two elements are equal are themselves equal. With this freedom we gain the hierarchy of $n$-types, allowing an even treatment of propositions, sets and groups. Where this freedom will come into its own is in mathematics and physics, but we also gain an advantage where we treat the symmetry groups of aspects of any situation.

One kind of reassuring *constraint* for a new language takes the form of the breadth of its application. I have suggested throughout the book that we understand the components of modal HoTT both through their appearance in language and thought and in their more refined appearance in mathematics. We saw, for instance, in Chap. 2 the commonality between the conjunction '*and*' of ordinary speech and the fibre bundles of mathematics, both being formed by the dependent sum construction. Chap. 4 saw discussion of invariance and change under forms of variation both in ordinary modal situations and as generalized in the mathematical treatment of monadic modalities. Then in Chap. 5 we discussed how visual imagination relies on a similar gluing to that employed in the higher spaces treated in current mathematics.

Elsewhere (Corfield 2017b) I described this phenomenon of the appearance of an idea realized in everyday thought and then refined mathematically as *vertical unity*, following the mathematician Alexandre Borovik. We find something of this vertical unity even when combining the components of modal HoTT: types, dependency, homotopy and modality. While the combination of spatial modalities with higher equivalence, or equivariance under higher symmetry, finds a major application in mathematical physics, as will be detailed elsewhere, we may see all of the components converge in the treatment of ordinary cognition. Take, for instance, the idea that carving out an event from an array of processes requires attention to its boundary, just as does the detection of an object from an array of substances. In the case of events, we saw in §2.6 how we pay great attention to this boundary in our depiction of *event nuclei*, where an interval of some activity culminates in the accomplishment of a change of state. But as we saw towards the end of §4.2.1, Cassirer had proposed a group-invariance account of the perception of entities, such as objects and tunes. Furthermore, such invariance arises within the kind of geometric setting described in Chap. 5, in which topological and Lie groups find their natural place. It should not be surprising, then, to find group-invariance and geometry appealed to also by current cognitive scientists studying the perception of events (see, for example, Shepard (1984) and Maguire et al. (2011)). Indeed, the latter paper provides a geometric rationale for the commonality between object and event individuation. Just as we distinguish an arm as a subpart of the human body due to the sudden change of curvature of the outline as it meets the torso, so any sudden curvature in a path is suggestive of either a collision or of internal decisions by an animate being. Such points in a path mark the unpredictability of the future and are thus of great interest to the perceiver.

Another form of constraint which acts to justify our new logic was described under the banner of *computational trinitarianism*. We saw here that a nexus of constructions simultaneously makes sense in type theory, in category theory and in programming. This confluence is very helpful, as the ways of thinking of each discipline can provide important insights when translated to the other corners of the *Trinity*. Where category theory has been slow to gain any purchase in philosophy, type theory will look far more familiar to those raised on first-order logic. In current mathematics, on the other hand, we have the reverse situation. Even those pessimistic about the prospects of HoTT for advancing mathematics itself rather than its philosophical treatment are committed users of category theory. And the third corner, programming, provides an excellent opportunity to test ideas in practice, where unforgiving machines are made to process effective calculi. We are starting to see this for mathematical proof and natural language processing. The *Trinity* comes as a bundle–adopt any one member and you adopt them all, even though you may be most at home in one corner.

If, as I propose, our type theory *is* taken up by philosophy, we should expect the programming theory of computer science to become more prevalent there. In view of their work already on calculi designed for the topics in metaphysics–time, event, process, and so on–quite possibly we will come to see computer scientists as doing a form of applied philosophy. It is certainly encouraging to see there the extensive use of types and monads, and, as remarked in Chap. 4, these are beginning to appear in linguistics, especially to deal with the pragmatic aspects of speech. Even so, some philosophers are sure to resist. Towards the conclusion of the paper we discussed in §2.6, in which he expresses his deep scepticism for Davidson's first-order treatment of events, Peter Hacker asks

> Even if, in some millennial logico-semantic paradise, a calculus were devised which mapped 'every difference and connection legitimately considered the business of a theory of meaning' on to a canonical notation, what would that show about our understanding of our native tongue? (Hacker 1982, p. 486)

Well, if indeed modal HoTT can furnish this paradise, so that computers could better operate on natural language semantics, then, together with its employment in physics, this would bring into closer proximity the four kinds of metaphysical notion given by Casati and Varzi, as outlined in §1.4. Recall these were: a common-sense notion; a philosophically refined notion, a scientifically refined notion and a psychological notion. The ensuing cross-fertilization of fields would surely enhance our philosophical imagination.

The expressive freedom of the modal aspect of the calculus is still very much under construction. Particularly interesting for philosophy will be the further development of temporal versions, and also probabilistic ones. We have only just begun to realize the promise of modal HoTT. A bright future awaits us.

# FURTHER READING

For any terms encountered in this book that the reader would like to see explained in more formal detail, it is worth consulting the $n$Lab wiki (https://ncatlab.org/nlab/show/HomePage). This is a tremendous resource with currently over 14,000 hyperlinked entries. It can only improve as time passes, and new contributors are always welcome.

The vast majority of the chosen texts listed below are open access.

## Category Theory

We are now blessed with a plethora of introductory texts. It is worth trying out different options. Readers with their varied backgrounds will find different texts more approachable.

- Emily Riehl, 2016. Category Theory in Context, Dover.
- Tom Leinster, 2014. Basic Category Theory, Cambridge University Press.
- Steve Awodey, 2010. Category Theory (2nd edition), Oxford University Press.
- Brendan Fong and David Spivak, 2019. An Invitation to Applied Category Theory: Seven Sketches in Compositionality, Cambridge University Press.

For the last of these there are

- Videos by the authors, http://math.mit.edu/dspivak/teaching/sp18/
- A companion online lecture course by John Baez which works through the book, https://www.azimuthproject.org/azimuth/show/Applied+Category+Theory+Course

## Type Theory

- Per Martin-Löf, The Collected Works, https://github.com/michaelt/martin-lof
- Erik Palmgren, 2014. Lecture Notes on Type Theory, http://staff.math.su.se/palmgren/lecturenotesTT.pdf
- Nicolo Gambino, 2009. Lectures on Dependent Type Theory, http://www1.maths.leeds.ac.uk/pmtng/Slides/dtt.html

### *Homotopy Type Theory*

The key textbook remains

- Univalent Foundations Program, 2014. Homotopy Type Theory: Univalent Foundations of Mathematics, http://homotopytypetheory.org/book/

Other resources:

- Steve Awodey, 2017. A Proposition is the (Homotopy) Type of its Proofs, https://arxiv.org/abs/1701.02024
- Daniel Grayson, 2017. An Introduction to Univalent Foundations for Mathematicians, https://arxiv.org/abs/1711.01477
- Andrej Bauer and Jaka Smrekar, 2019. Homotopy (Type) Theory, https://github.com/andrejbauer/homotopy-type-theory-course
- Egbert Rijke, 2018. Introduction to Homotopy Type Theory, http://www.andrew.cmu.edu/user/erijke/hott/

### *Modal Type Theory*

- Dan Licata and Felix Wellen, 2018. Synthetic Mathematics in Modal Dependent Type Theories, http://dlicata.web.wesleyan.edu/pubs/lsr17multi/him-tutorial.pdf (videos available)

### *The Relationship Between Type Theory and Category Theory*

- nLab: Relation Between Type Theory and Category Theory
- Mike Shulman, 2018. Homotopical Trinitarianism: A Perspective on Homotopy Type Theory, https://home.sandiego.edu/ shulman/papers/trinity.pdf

### *Miscellaneous*

- Research-level talks, Homotopy Type Theory Electronic Seminar Talks, https://www.uwo.ca/math/faculty/kapulkin/seminars/hottest.html
- Papers in synthetic mathematics, https://ncatlab.org/nlab/show/mathematics+presented+in+homotopy+type+theory
- Type theory for computing resources, https://github.com/jozefg/learn-tt

# REFERENCES

Abramsky, S. and Coecke, B. 2008. 'Categorical Quantum Mechanics' in *Handbook of Quantum Logic and Quantum Structures*, Elsevier, 261–324, arXiv:0808.1023.

Anel, M., Biedermann, G., Finster, E. and Joyal, A. 2017. 'A Generalized Blakers–Massey Theorem,' arXiv:1703.09050.

Anel, M. and Joyal, A. 2019. 'Topo-logie,' to appear in G. Catren and M. Anel (eds) *New Spaces for Mathematics and Physics: Formal and Conceptual Reflections*.

Arp R., Smith, B. and Spear, A. 2015. *Building Ontologies with Basic Formal Ontology*, MIT Press.

Asudeh, A. and Giorgolo, G. 2016. 'Perspectives,' *Semantics and Pragmatics*, 9(21).

Atiyah, M. 2003. 'What is Geometry?' in C. Pritchard (ed.) *The Changing Shape of Geometry: Celebrating a Century of Geometry and Geometry Teaching*, Cambridge University Press, 24–30.

Awodey, S. 2014. 'Structuralism, Invariance, and Univalence,' *Philosophia Mathematica*, 22(1): 1–11.

Awodey, S. and Kishida, K. 2008. 'Topology and Modality: The Topological Interpretation of First-order Modal Logic,' *The Review of Symbolic Logic*, 1(2): 146–66.

Awodey, S. and Kishida, K. 2012. 'Topological Completeness of First-Order Modal Logic,' *Advances in Modal Logic*, 9: 1–17.

Bach, E. 1986. 'The Algebra of Events,' *Linguistics and Philosophy*, 9(1): 5–16.

Baez, J. and Biamonte, J. 2018. *Quantum Techniques for Stochastic Mechanics*, World Scientific, Singapore.

Baez, J. and Dolan, J. 2001. 'From Finite Sets to Feynman Diagrams' in B. Engquist and W. Schmid (eds) *Mathematics Unlimited—2001 and Beyond, Vol. 1*, Springer, Berlin, 29–50, arXiv:0004133.

Baez, J. and Schreiber, U. 2005. 'Higher Gauge Theory' in A. Davydov et al, (eds) *Categories in Algebra, Geometry and Mathematical Physics*, Contemp. Math. 431, AMS, Providence, Rhode Island, 2007, 7–30, arXiv:0511710.

Baez, J. and Stay, M. 2010. 'Physics, Topology, Logic and Computation: A Rosetta Stone' in B. Coecke (ed.) *New Structures for Physics*. Lecture Notes in Physics, vol 813. Springer, Berlin, Heidelberg, 95–172.

Balbiani, P., Goranko, V. and Sciavicco, G. 2011. 'Two-sorted Point-Interval Temporal Logics,' *Proc. of the 7th International Workshop on Methods for Modalities (M4M7)*, Electronic Notes in Theoretical Computer Science, 278, Elsevier, 31–45.

Benacerraf, P. 1965. 'What Numbers Could Not Be' in P. Benacerraf and H. Putnam (eds). *Philosophy of Mathematics: Selected Readings*, Cambridge University Press, 2nd edition, 1983, 272–94.

Ben-Zvi, D. 2014. Blog comment. https://www.math.columbia.edu/~woit/wordpress/?p=7114&cpage=1#comment-214353

Black, M. 1952. 'The Identity of Indiscernibles,' *Mind*, 61(242): 153–64.

Boghossian, P. 2011. 'Williamson on the *A Priori* and the Analytic,' *Philosophy and Phenomenological Research*, 82(2): 488–97.

Borger, J. 2009. Blog comment. https://golem.ph.utexas.edu/category/2009/02/lakatos_as_dialectical_realist.html#c022225

Brady, G. and Trimble, T. 2000a. 'A Categorical Interpretation of C. S. Peirce's Propositional Logic Alpha,' *Journal of Pure and Applied Algebra*, 149: 213–39.

Brady, G. and Trimble, T. 2000b. 'A String Diagram Calculus for Predicate Logic and C. S. Peirce's System Beta,' preprint, https://ncatlab.org/nlab/files/BradyTrimbleString.pdf.

Brandom, R. 1994. *Making it Explicit*, Harvard University Press.

Brandom, R. 2000. *Articulating Reasons*, Harvard University Press.

Brandom, R. 2010. *Between Saying and Doing*, Oxford University Press.

Brandom, R. 2015. *From Empiricism to Expressivism*, Harvard University Press.

Carchedi, D. 2013. 'Higher Orbifolds and Deligne-Mumford Stacks as Structured Infinity Topoi,' arXiv:1312.2204.

Carnap, R. 1932. 'The Elimination of Metaphysics Through Logical Analysis of Language,' *Erkenntnis*, 3: 60–81.

Casati, R. and Varzi, A. 2008. 'Understanding Events' in T. Shipley and J. Zacks (eds) *From Perception to Action*, New York, Oxford University Press, 31–54.

Casati, R. and Varzi, A. 2015. 'Events,' in E. N. Zalta (ed.) *The Stanford Encyclopedia of Philosophy* (Winter 2015 Edition).

Cassirer, E. 1925. *Language and Myth*, 1925, tr. S. Langer, Dover, 1953.

Cassirer, E. 1944. 'The Concept of Group and the Theory of Perception, *Philosophy and Phenomenological Research*, 5(1): 1–36.

Cassirer, E. 1957. *The Philosophy of Symbolic Forms*, iii: *The Phenomenology of Knowledge*, Yale University Press.

Chatzikyriakidis, S. and Luo, Z. 2017. 'On the Interpretation of Common Nouns: Types vs Predicates' in S. Chatzikyriakidis and Z. Luo (eds) *Modern Perspectives in Type-Theoretical Semantics*, Studies in Linguistics and Philosophy Springer, 43–70.

Coecke, B., Sadrzadeh, M. and Clark, S. 2010. 'Mathematical Foundations for a Compositional Distributional Model of Meaning, *Lambek Festschrift: Linguistic Analysis*, 36: 345–84, arXiv:1003.4394.

Collier, J. 2004. 'Self-Organization, Individuation and Identity,' *Revue internationale de philosophie* 228: 151–72.

Collingwood, R. 1939. *An Autobiography*. Oxford: Clarendon Press.

Collingwood, R. 1940. *An Essay on Metaphysics*, Oxford: Clarendon Press.

Conrad, B. *n.d.* 'Math 121. Uniqueness of Algebraic Closure,' handout, <http://math.stanford.edu/~conrad/121Page/handouts/algclosure.pdf>.

Corfield, D. 2003. *Towards a Philosophy of Real Mathematics*, Cambridge University Press.

Corfield, D. 2004. 'Mathematical Kinds, or Being Kind to Mathematics,' *Philosophica*, 74: 30–54.

Corfield, D. 2010. 'Lautman et la réalité des mathématiques,' *Philosophiques*, 37(1): 95–109, in English as 'Lautman and the Reality of Mathematics,' http://philsci-archive.pitt.edu/9210/.

Corfield, D. 2012. 'Narrative and the Rationality of Mathematical Practice' in A. Doxiadis and B. Mazur (eds) *Circles Disturbed: The Interplay of Mathematics and Narrative*, Princeton University Press, 244–80.

Corfield, D. 2017a. 'Expressing "The Structure of" in Homotopy Type Theory,' *Synthese*, https://doi.org/10.1007/s11229-017-1569-7.

Corfield, D. 2017b. 'Homotopy Type Theory and the Vertical Unity of Concepts in Mathematics' in E. de Freitas, N. Sinclair and A. Coles (eds) *What is a Mathematical Concept?* Cambridge University Press, 125–42.

Corfield, D. 2017c. 'Reviving the Philosophy of Geometry' in E. Landry (ed.) *Categories for the Working Philosopher*, Oxford University Press, 18–36.

Corfield, D. 2017d. 'Duality as a Category-Theoretic Concept,' *Studies in History and Philosophy of Modern Physics* 59: 55–61.

Dancy, J. 2017. 'Moral Particularism' in Edward N. Zalta (ed.) *The Stanford Encyclopedia of Philosophy* (Winter 2017 Edition).

Davidson, D. 2001. *Essays on Actions and Events*: 2nd edition, Oxford: Clarendon Press.

de Swart, H. 1998. 'Aspect Shift and Coercion,' *Natural Language and Linguistic Theory* 16, 347–85.

DiFrisco, J. 2018. 'Biological Processes: Criteria of Identity and Persistence,' in D. Nicholson and J. Dupré (eds), *Everything Flows: Towards a Processual Philosophy of Biology*, Oxford University Press, 76–95.

Domski, M. and Dickson, M. (eds). 2010. *Discourse on a New Method: Reinvigorating the Marriage of History and Philosophy of Science*, Open Court Publishing Company.

Dummett, M. 1977. *Elements of Intuitionism*, Oxford: Clarendon Press.

Dummett, M. 1991. *The Logical Basis of Metaphysics*, Harvard University Press.

Einstein, A. 1934. 'On The Method of Theoretical Physics,' *Philosophy of Science* 1(2): 163–9.

Fairtlough, M. and Mendler, M. 2002. 'On the Logical Content of Computational Type Theory: A Solution to Curry's Problem' in P. Callaghan, Z. Luo and J. McKinna (eds) *Types for Proofs and Programs*, Springer (LNCS 2277), 63–78.

Field, H. 1980. *Science without Numbers*, Oxford University Press.

Fine, K. 1972. 'In so many Possible Worlds,' *Notre Dame Journal of Formal Logics* 13: 516–20.

Fiorenza, D., Sati, H. and Schreiber, U. 2019. 'The Rational Higher Structure of M-theory,' *Proceedings of the LMS EPSRC Durham Symposium: Higher Structures in M-Theory*, August 2018, Fortschritte der Physik, arXiv:1903.02834.

Fong, B., Myers, D. and Spivak, D. 2018. Behavioral Mereology, arXiv:1811.00420.

Freed, D. and Teleman, C. 2018. *Topological Dualities in the Ising Model*. arXiv:1806.00008.

Friedman, M. 2000. *The Parting of the Ways: Carnap, Cassirer, and Heidegger*, Open Court Publishing Company.

Friedman, M. 2001. *Dynamics of Reason*, CSLI Publications.

Friedman, M. 2002. 'Kant, Kuhn, and the Rationality of Science,' *Philosophy of Science* 69(2): 171–90.

Gaboardi, M., Katsumata, S., Orchard, D., Breuvart, F. and Uustalu, T. 2016. 'Combining Effects and Coeffects via Grading'. *ICFP 2016 Proceedings of the 21st ACM SIGPLAN International Conference on Functional Programming*, 476–89.

Gaitsgory, D. and Lurie, J. 2019. *Weil's Conjecture for Function Fields, Volume I (AMS-199)*, Annals of Mathematics Studies, 360, Princeton University Press.

Galison, P. 1997. *Image and Logic: A Material Culture of Microphysics*, University of Chicago Press.

Garson, J. 2018. 'Modal Logic,' in Edward N. Zalta (ed.) *The Stanford Encyclopedia of Philosophy* (Fall 2018 Edition).

Giorgolo, G. and Asudeh, A. 2012. '$\langle M, \eta, \star \rangle$ Monads for Conventional Implicatures' in A. Guevara, A. Chernilovskaya and R. Nouwen, (eds.) *Proceedings of Sinn and Bedeutung*, 16, MIT Working Papers in Linguistics.

Giovanelli, M. 2013. 'Talking at Cross-Purposes: How Einstein and The Logical Empiricists Never Agreed on What They Were Disagreeing About,' *Synthese* 190(17): 3819–63.

Goble, L. 1970. 'Grades Of Modality,' *Logique et Analyse*, 13: 323–34.

Goldblatt, R. 1992. *Logics of Time and Computation*: 2nd Edition, CSLI Publications.

Goldman, A. 2007. 'A Program for "Naturalizing" Metaphysics, with Application to the Ontology of Events,' *The Monist* 90: 457–79.

González de Prado Salas, J., de Donato Rodríguez, X. and Zamora Bonilla, J. 2017. 'Inferentialism, Degrees of Commitment, and Ampliative Reasoning,' *Synthese*: 1–19, https://doi.org/10.1007/s11229-017-1579-5.

Goodman, N. 1951. *The Structure of Appearance*, Harvard University Press.

Goodman, N. 1955. *Fact, Fiction, and Forecast*, Harvard University Press.

Greenlees, J. and May, J. 1995. 'Equivariant Stable Homotopy Theory' in I. James (ed.) *Handbook of Algebraic Topology*, Elsevier, 279–325.

Grice P. 1989. *Studies in the Way of Words*, Harvard University Press.

Grothendieck, A. 1983. 'Pursuing Stacks,' Letter to D. Quillen in G. Maltsiniotis, M. Künzer and B. Toen (eds) *Documents Mathématiques*, Society Mathematics Paris, France.

Grothendieck, A. and Verdier, J. 1972. *Théorie des topos et cohomologie étale des schémas, Tome 1: Théorie des topos* in M. Artin, A. Grothendieck, and J. L.Verdier (eds) *Lecture Notes in Mathematics*, 269, Berlin: Springer-Verlag, SGA 4.

Hacker P. 1982. 'Events, Ontology and Grammar,' *Philosophy*, 57(222): 477–86.

Hacker, P. 2001. 'On Strawson's Rehabilitation of Metaphysics,' in *Wittgenstein: Connections and Controversies*, Oxford University Press, 345–69.

Hacker, P. 2010. 'A Normative Conception of Necessity: Wittgenstein on Necessary Truths of Logic, Mathematics and Metaphysics,' *Publications of the Austrian Ludwig Wittgenstein Society—New Series*, 14: 13–33.

Halpern, J., Harper, R., Immerman, N., Kolaitis, P., Vardi, M. and Vianu, V. 2001. 'On the Unusual Effectiveness of Logic in Computer Science,' *Bulletin of Symbolic Logic*, 7(2): 213–36.

Harper, R. 2011. 'The Holy Trinity,' blog post on The Existential Type, https://existentialtype. wordpress.com/2011/03/27/the-holy-trinity/.

Harré, R., Aronson, E. and Way, J. 1994. *Realism Rescued: How Scientific Progress Is Possible*, Gerald Duckworth and Co. Ltd.

Hayaki, R. 2003. 'Actualism and Higher-Order Worlds,' *Philosophical Studies* 115: 149–78.

Heidegger, M. 1967. 'Modern Science, Metaphysics, and Mathematics' in D. F. Krell (ed.) *Basic Writings*, London: Routledge and Kegan Paul, 1978, 271–305.

Heis, J. 2011. 'Ernst Cassirer's Neo-Kantian Philosophy of Geometry,' *British Journal for the History of Philosophy* 19(4): 759–94.

Henriques, A. 2010. Comment on 'What's a groupoid? What's a good example of a groupoid?' (http://mathoverflow.net/a/1161).

Hilbert, D. 1927. 'The Foundations of Mathematics,' in J. van Heijenoort (ed.) 2002. *From Frege to Gödel*, Harvard University Press.

Hou, K.-B. (Favonia), Finster, E., Licata, D. and Lumsdaine, P. 2016. *A mechanization of the Blakers-Massey connectivity theorem in Homotopy Type Theory*. arXiv.1605.03227.

Hou, K.-B. (Favonia) and Harper, R. 2018. 'Covering Spaces in Homotopy Type Theory'. *22nd International Conference on Types for Proofs and Programs (TYPES 2016), Leibniz International Proceedings in Informatics* 11: 1–16.

Huerta, J., Sati, H. and Schreiber, U. 2018. *Real ADE-equivariant (co)homotopy and Super M-branes*. arXiv:1805.05987.

Jaffe, A. and Quinn, F. 1993. ' "Theoretical mathematics": Toward a cultural synthesis of mathematics and theoretical physics,' arXiv:math/9307227.

Johnstone, P. 2002. *Sketches of an Elephant: A Topos Theory Compendium*. Oxford Logic Guides, Oxford: Clarendon Press.

Keeping, J. 2014. 'The Time Is Out of Joint: A Hermeneutic Phenomenology of Grief,' *Symposium* 18(2): 233–55.

Keränen, J. 2001. 'The Identity Problem for Realist Structuralism'. *Philosophia Mathematica* 9: 308–30.

Keränen, J. 2006. 'The Identity Problem for Realist Structuralism II: A Reply to Shapiro' in MacBride (ed.) *Identity and Modality*, Oxford: Clarendon Press, 146–63.

Kishida, K. 2017. 'Categories and Modalities' in E. Landry (ed.) *Categories for the Working Philosopher*, Oxford University Press, 163–222.

Kripke, S. 1980. *Naming and Necessity*, Harvard University Press.

Ladyman, J. and Presnell, S. 2015. 'Identity in Homotopy Type Theory, Part I: The Justification of Path Induction,' *Philosophia Mathematica* 23: 386–406.

Ladyman, J. and Presnell, S. 2016. 'Does Homotopy Type Theory Provide a Foundation for Mathematics?,' *British Journal for the Philosophy of Science* 69: 377–420.

Ladyman, J. and Presnell, S. 2017. 'Identity in Homotopy Type Theory, Part II: The Conceptual and Philosophical Status of Identity in HoTT,' *Philosophia Mathematica*, 25: 210–45.

Ladyman, J. and Ross, D. 2007. *Everything Must Go*, Oxford University Press.

Lambek, J. and Scott, P. 1986. *Introduction to Higher-Order Categorical Logic*, Cambridge University Press.

Landry, E. (ed.). 2017. *Categories for the Working Philosopher*, Oxford University Press.

Lassiter, D. 2017. *Graded Modality: Qualitative and Quantitative Perspectives*, Oxford University Press.

Lautman, A. 2006. *Les mathématiques les idées et le réel physique*, Zalamea, F. (ed.), Vrin, Paris.

Lawvere, F W. 1969. 'Adjointness in Foundations,' *Dialectica*, 23: 281–96. Republished in *Reprints in Theory and Applications of Categories*, 16: 1–16, 2006.

Lawvere, W. 1970. 'Quantifiers and Sheaves,' *Actes du congres international des mathematiciens, Nice*: 329–34.

Lawvere, W. 1973. 'Metric Spaces, Generalized Logic and Closed Categories,' *Rendiconti del Seminario Matematico e Fisico di Milano XLIII*, 135–66. Republished in *Reprints in Theory and Applications of Categories* 1 (2002): 1–37.

Lawvere, W. 1986. 'Taking Categories Seriously,' *Revista Colombiana de Matematicas, XX*, 147–78. Republished in *Reprints in Theory and Applications of Categories* 8: 1–24 (2005).

Lawvere, W. 1991. 'Some Thoughts on the Future of Category Theory' in A. Carboni et al. (eds) *Category Theory*, Springer, 1–13.

Lawvere, W. 1994. 'Cohesive Toposes and Cantor's "lauter Einsen",' *Philosophia Mathematica* 3(2): 5–15.

Lawvere, W. 2000. 'Adjoint Cylinders,' Categories mailing list comment, http://permalink.gmane.org/gmane.science.mathematics.categories/1683.

Lawvere, W. 2007. 'Axiomatic Cohesion,' *Theory and Applications of Categories* 19(3): 41–9.

Leinster, T. and Cobbald, C. 2012. 'Measuring Diversity: The Importance of Species Similarity for a Way to Count Biodiversity,' *Ecology* 93(3): 477–89.

Lewis, D. 1973. *Counterfactuals*, Oxford: Blackwell Publishing.

Lewis, D. 1986. *On the Plurality of Worlds*, Oxford: Blackwell Publishing.

Licata, D. and Shulman, M. 2013. 'Calculating the Fundamental Group of the Circle in Homotopy Type Theory' in *LICS 2013: Proceedings of the Twenty-Eighth Annual ACM/IEEE Symposium on Logic in Computer Science*.

Licata, D. and Shulman, M. 2016. 'Adjoint Logic with a 2-category of Modes' in *Logical Foundations of Computer Science*, Springer, 219–35.

Licata, D., Shulman, M. and Riley, M. 2017. 'A Fibrational Framework for Substructural and Modal Logics (extended version)' in *Proceedings of 2nd International Conference on Formal Structures for Computation and Deduction (FSCD 2017)* (doi: 10.4230/LIPIcs.FSCD.2017.25).

Lindström, S. and Segerberg, K. 2007. 'Modal Logic and Philosophy' in P. Blackburn, J. van Benthem and F. Wolter (eds) *Handbook of Modal Logic*, Elsevier, 1149–1214.

Lowe, E. J. 1989. 'Impredicative Identity Criteria and Davidson's Criterion of Event Identity,' *Analysis* 49(4): 178–81.

Luo, Z. 2012. 'Common Nouns as Types' in D. Béchet, and A. Dikovsky (eds) *Logical Aspects of Computational Linguistics*, Lecture Notes in Computer Science, 7351, Springer, 173–85.

Lurie, J. 2009a. *Higher Topos Theory*, Princeton University Press.

Lurie, J. 2009b. 'Derived Algebraic Geometry V: Structured Spaces,' arXiv:0905.0459.

MacBride, F. (ed.). 2006. *Identity and Modality*, Oxford: Clarendon Press.

Mac Lane, S. 1998. *Categories for the Working Mathematician*, Springer.

Maguire, M. Brumberg, M., Ennis, M. and Shipley, T. 2011, 'Similarities in Object and Event Segmentation: A Geometric Approach to Event Path Segmentation,' *Spatial Cognition and Computation*, 11(3): 254–79.

Mancosu, P. (ed.). 2008. *The Philosophy of Mathematical Practice*, Oxford University Press.

Marquis, J.-P. 2008. *From a Geometric Point of View: A Study of the History and Philosophy of Category Theory*, Springer.

Martin-Löf P. 1984. *Intuitionistic Type Theory, Studies in Proof Theory*, Napoli: Bibliopolis.

Maršík, J. and Amblard, M. 2016. 'Introducing a Calculus of Effects and Handlers for Natural Language Semantics,' *Proceedings of the 20th and 21st International Conferences on Formal Grammar*, 9804, Springer-Verlag, 257–72.

Mazel-Gee, A. 2016. *Goerss–Hopkins Obstruction Theory via Model 1-Categories*, PhD thesis.

McLarty, C. 1990. 'The Uses and Abuses of the History of Topos Theory,' *British Journal for the Philosophy of Science*, 41(3): 351–75.

McLarty C. 2007. 'The Last Mathematician from Hilbert's Göttingen: Saunders Mac Lane as Philosopher of Mathematics,' *British Journal for the Philosophy of Science*, 58(1): 77–112.

McLarty, C. 2008. ' "There is No Ontology Here:" Visual and Structural Geometry in Arithmetic' in P. Mancosu (ed.) *The Philosophy of Mathematical Practice*, Oxford University Press, 370–406.

Melliès, P. and Zeilberger, N. 2016. A Bifibrational Reconstruction of Lawvere's Presheaf Hyperdoctrine arXiv:1601.06098.

Moens, M. and Steedman, M. 1988. 'Temporal Ontology and Temporal Reference,' *Computational Linguistics*, 14(2): 15–28.

Montague, R. 1974. 'Formal Philosophy' in *Selected Papers of Richard Montague*, New Haven: Yale University Press.

MSRI, 2014. 'Geometric Representation Theory' Programme announcement, https://www.msri.org/programs/276.

Nagel, E. 1939. 'The Formation of Modern Conceptions of Formal Logic in the Development of Geometry,' *Osiris* 7: 142–223.

Nicholson, D. and Dupré, J. (eds). 2018. Everything Flows: Towards a Processual Philosophy of Biology, Oxford University Press.

nLab, 'general covariance,' https://ncatlab.org/nlab/show/general+covariance.

Nodelman, U. and Zalta, E. 2014. 'Foundations for Mathematical Structuralism,' *Mind* 123/489: 39–78.

Paseau, A. 2005. 'What the Foundationalist Filter Kept Out,' *Studies in History and Philosophy of Science Part A* 36(1): 191–201.

Pearl, J. 2009. Causality: *Models, Reasoning, and Inference*: 2nd edition, Cambridge University Press.

Peirce, C. 1906. 'Prolegomena to an Apology for Pragmaticism,' *The Monist* 16: 492–546.

Peirce, C. 1992. *Reasoning and the Logic of Things*, K. Kettner (ed.), with an introduction by K. Ketner and H. Putnam, Harvard University Press.

Polanyi, M. 1958. *Personal Knowledge: Towards a Post-critical Philosophy* University of Chicago Press.

Quine, W. 1986. *Philosophy of Logic*: 2nd edition, Harvard University Press, Englewood Cliffs (NJ): Prentice-Hall.

Quine, W. and Ullian, J. 1970. *The Web of Belief*, New York: Random House.

Ramsey, F. 1931. 'Philosophy' in R. B. Braithwaite (ed.), *The Foundations of Mathematics and Other Logical Essays by Frank Plumpton Ramsey*, London: Harcourt, Brace.

Ranta, A. 1991. 'Constructing Possible Worlds,' *Theoria* 57(1–2): 77–99.

Ranta, R. 1994. *Type-Theoretic Grammar*, Oxford: Clarendon Press.

Ranta, A. 2015. 'Constructive Type Theory' in S. Lappin and C. Fox (eds) *The Handbook of Contemporary Semantic Theory*, 2nd edition, Wiley Blackwell, 345–80.

Redding, P. 2007. *Analytic Philosophy and the Return of Hegelian Thought*, Cambridge University Press.

Reed, J. 2009. 'A Judgmental Deconstruction of Modal Logic,' http://www.cs.cmu.edu/jcreed/papers/jdml.pdf

Resnik, M. 1997. *Mathematics as a Science of Patterns*, Oxford: Clarendon Press.

Rezk, C. 2014. 'Global Homotopy Theory and Cohesion,' preprint available at https://faculty.math.illinois.edu/ rezk/global-cohesion.pdf.

Rezk, C. 2015. 'Proof of the Blakers-Massey Theorem,' preprint available at https://faculty.math.illinois.edu/rezk/freudenthal-and-blakers-massey.pdf

Rota, G.-C. 1991. 'The Pernicious Influence of Mathematics upon Philosophy,' *Synthese*, 88(2): 165–78.

Rundle, B. 1979. *Grammar in Philosophy*, Oxford University Press.

Rundle, B. 1983. 'Conjunctions: Meaning, Truth and Tone,' *Mind*, 92(367): 386–406.

Russell, B. 1905. 'On Denoting,' *Mind*, 14: 479–93.

Russell, B. 1914. 'Logic as the Essence of Philosophy,' *Lecture II of Our Knowledge of the External World, Chicago and London*, The Open Court Publishing Company.

Ryckman, T. 2005. *The Reign of Relativity*, Oxford University Press.

Ryle, G. 1949. *The Concept of Mind*, London: Hutchinson.

Scedrov, A. 1986. 'Diagonalization of Continuous Matrices as a Representation of Intuitionistic Reals,' *Annals of Pure and Applied Logic*, 30: 201–6.

Scholz, E. 2005. 'Philosophy as a Cultural Resource and Medium of Reflection for Hermann Weyl,' *Revue de Synthèse*, 126: 331–51.

Scholz, E. 2011. 'H. Weyl's and E. Cartan's Proposals for Infinitesimal Geometry in the Early 1920s,' *Boletim da Sociedada Portuguesa de Matematica* Numero Especial A: 225–45.

Schreiber, U. 2013. 'Differential Cohomology in a Cohesive Infinity-topos,' arXiv/1310.7930, latest version at https://ncatlab.org/schreiber/show/differential+cohomology+in+a+cohesive+topos

Schreiber, U. 2014a. 'Quantization via Linear homotopy types,' arXiv:1402.7041.

Schreiber, U. 2014b, nForum discussion comment, http://nforum.ncatlab.org/discussion/2084/higher-geometry/?Focus=49931#Comment_Schreiber49931

Schreiber, U. (forthcoming) 'Higher Prequantum Geometry' in G. Catren and M. Anel (eds.) *New Spaces for Mathematics and Physics*, arXiv:1601.05956

Schreiber, U. and Shulman, M. 2014. 'Quantum Gauge Field Theory in Cohesive Homotopy Type Theory,' arXiv:1408.0054.

Schreiber, U. et al. 2014. 'Explaining the Point of HoTT on FOM'. Google+ discussion archived at https://github.com/DavidMichaelRoberts/Sandbox/blob/master/Schreiber_Gplus_post.md

Schultz, P. and Spivak, D. 2017. Temporal Type Theory: A Topos-theoretic Approach to Systems and Behavior', arXiv:1710.10258.

Sellars, W. 1953. 'Inference and Meaning,' *Mind* 62: 313–38.

Shapiro, S. 1997. *Philosophy of Mathematics: Structure and Ontology*, Oxford University Press.

Shepard, R. 1984. 'Ecological Constraints on Internal Representation: Resonant Kinematics of Perceiving, Imagining, Thinking, and Dreaming,' *Psychological Review* 91: 417–47.

Shulman, M. 2013. 'From Set Theory to Type Theory,' online article, https://golem.ph.utexas.edu/category/2013/01/from_set_theory_to_type_theory.html

Shulman, M. 2017. 'Homotopy Type Theory: A synthetic Approach to Higher Equalities' in E. Landry (ed.) *Categories for the Working Philosopher*, Oxford University Press, arXiv:1601.05035.

Shulman, M. 2018a. 'Brouwer's Fixed-point Theorem in Real-cohesive Homotopy Type Theory, arXiv:1509.07584 and *Mathematical Structures in Computer Science* 28(6), June 2018: 856–941.

Shulman, M. 2018b. *Homotopical Trinitarianism: A Perspective on Homotopy Type Theory, a Talk at the Homotopy Type Theory (MRC) Special Session at the 2018 Joint Mathematics Meetings*, https://home.sandiego.edu/ shulman/papers/trinity.pdf.

Shulman, M. Forthcoming. 'Homotopy Type Theory: The Logic of Space,' to appear in G. Catren and M. Anel (eds) *New Spaces for Mathematics and Physics: Formal and Conceptual Reflections*, arXiv:1703.03007.

Smith, B. 2005. 'Against Fantology' in J. C. Marek and M. E. Reicher (eds), *Experience and Analysis*, Kirchberg am Wechsel, 153–70.

Stanley, J. 2007. *Language in Context*, Oxford University Press.

Steedman, M. 2012. 'Computational Linguistics,' in R. Binnick (ed.) *The Oxford Handbook of Tense and Aspect*, Oxford University Press, 102–20.

Strawson, P. 1950. 'On Referring,' *Mind* 59: 320–44.

Strawson, P. 1959. *Individuals: An Essay in Descriptive Metaphysics*, Routledge.

Strawson, P. 1964. 'Identifying Reference and Truth-Values,' *Theoria* Vol XXX, 96–118, reprinted as Chap. 4 of Strawson, P. F. 1971. *Logico-Linguistic Papers*, London: Methuen.

Sundholm, G. 1986. 'Proof Theory and Meaning' in D. Gabbay and F. Guenthner (eds), *Handbook of Philosophical Logic—Volume III*, Reidel: Kluwer, 471–506.

Tanaka, R., Mineshima, K. and Bekki, D. 2015. 'Factivity and Presupposition in Dependent Type Semantics'. Conference talk at TYTLES: TYpe Theory and LExical Semantics, Barcelona: https://www.lirmm.fr/tytles/Articles/Tanaka.pdf.

Tanaka, R., Mineshima, K. and Bekki, D. 2017. 'Factivity and Presupposition in Dependent Type Semantics,' *Journal of Language Modelling* 5(2): 385–420.

Tappenden, J. 1995. 'Extending Knowledge and "Fruitful Concepts:" Fregean Themes in the Foundations of Mathematics Jamie Tappenden,' *Nous* 29(4): 427–67.

Toën, B. 2014. 'Derived Algebraic Geometry,' arXiv:1401.1044.

Torretti, R. 1978. *Philosophy of Geometry from Riemann to Poincaré*, Springer.

Tsementzis, D. 2017. 'Univalent Foundations as Structuralist Foundations,' *Synthese* 194(9): 3583–617.

UFP (Univalent Foundations Program). 2014. 'Homotopy Type Theory: Univalent Foundations of Mathematics'. http://homotopytypetheory.org/book/.

van Fraassen, B. 1989. *Laws and Symmetry*, Oxford: Clarendon Press.

van Fraassen, B. 2002. *The Empirical Stance*, Yale University Press.

Vendler, Z. 1965. 'The Relevance of Linguistics to Philosophy: Comments,' *The Journal of Philosophy* 62(20): 602–5.

Vendler, Z. 1967a. *Linguistics in Philosophy*, Cornell University Press.

Vendler, Z. 1967b. 'Causal Relations,' *The Journal of Philosophy* 64(21): 704–13.

Venema, Y. 1991. 'A Modal Logic for Chopping Intervals'. *Journal of Logic and Computation*, 1(4): 453–76.

Voevodsky, V. 2014. 'The Origins and Motivations of Univalent Foundations: A Personal Mission to Develop Computer Proof Verification to Avoid Mathematical Mistakes,' https://www.ias.edu/ideas/2014/voevodsky-origins.

Walsh, P. 2017. 'Categorical Harmony and Path Induction,' *The Review of Symbolic Logic* 10(2): 301–21.

Wellen, F. 2018. 'Cartan Geometry in Modal Homotopy Type Theory,' arXiv:1806.05966.

Wellwood, A., Hespos, S. and Rips, L. 2017. 'The Object: Substance:: Event: Process Analogy' in T. Lombrozo, J. Knobe and S. Nichols (eds) *Oxford Studies in Experimental Philosophy*, Volume 2, Oxford University Press, 183–212.

Weyl, H. 1932. 'The Open World: Three Lectures on the Metaphysical Implications of Science'. Reprinted as Chap. 4 of P. Pesic (ed.) *Mind and Nature: Selected Writings on Philosophy, Mathematics, and Physics*, 34–82.

Wiggins, D. 1980. *Sameness and Substance*, Oxford: Blackwell Publishing.

Williamson, T. 2003. 'Everything,' *Philosophical Perspectives* 17(1): 415–65.

Williamson, T. 2013. *Modal Logic as Metaphysics*. Oxford University Press.

Wilson, M. 2006. *Wandering Significance*, Oxford University Press.

Wittgenstein, L. 1956. *Remarks on the Foundations of Mathematics*, G. H. von Wright, R. Rhees, G. E. M. Anscombe (eds.), Oxford: Basil Blackwell.

Xue, T., Luo, Z. and Chatzikyriakidis, S. 2018. 'Propositional Forms of Judgemental Interpretations,' *Fifth Workshop on Natural Language and Computer Science (NLCS18), Oxford*.

# INDEX